CONFERENCE SERIES

Series Editors: K. Habitzel, T. D. Märk, B. Stehno

universität
innsbruck

*i*up • *innsbruck* university press

www.uibk.ac.at/iup

© 2005 *innsbruck* university press
2nd edition
All rights reserved.

Vizerektorat für Forschung
Leopold-Franzens-Universität Innsbruck
Christoph-Probst-Platz, Innrain 52
A-6020 Innsbruck
www.uibk.ac.at/iup

Book editors: Armin Hansel, Tilmann Märk
Cover design: Carmen Drolshagen
Produced: Books on Demand GmbH

ISBN: 3-901249-78-8

2nd International Conference on Proton Transfer Reaction Mass Spectrometry and Its Applications

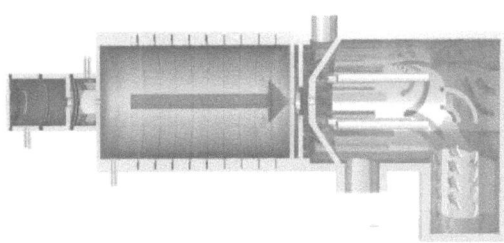

Contributions

Editors:

Armin Hansel
Tilmann Märk

Institut für Ionenphysik
der Leopold-Franzens-Universität Innsbruck
Technikerstr. 25
6020 Innsbruck, Austria

Obergurgl, Austria
January 29 – February 3, 2005

Acklowledgements

This conference was supported by the following companies:

PFEIFFER VACUUM Austria

IONICON ANALYTIK GmbH, Innsbruck, Austria

IONIMED ANALYTIK GmbH, Innsbruck, Austria

2nd International Conference on Proton Transfer Reaction Mass Spectrometry and Its Applications

Local Organizing Committee:

Armin Hansel (Chairman)
Tilmann Märk
Armin Wisthaler
Christine Oberhofer
Wolfgang Grabmer
Wolfgang Sailer

Institut für Ionenphysik
Innsbruck

Medieninhaber:
Armin Hansel, Institut für Ionenphysik
der Leopold-Franzens-Universität Innsbruck
Technikerstr. 25, 6020 Innsbruck, Austria

Foreword

Over the past few decades the need for quantitative and fast determination of a variety of organic compounds in complex air matrices at ultra low concentrations has continuously pushed the limits of analytical chemistry. For a wide variety of real-world problems mass spectrometry has provided unique and competitive solutions. In conventional mass spectrometry (MS) electron impact is used to ionise organic compounds. Unfortunately, this technique has a major drawback; instead of forming a single ionised species, many molecules break down into smaller fragment ions, which can result in a single compound giving rise to a complex "mass spectral pattern". With a mixture of several compounds entering the MS detector simultaneously, the final mass spectrum may be so complex that interpretation and quantification become difficult, if not impossible. The traditional solution to this problem has been to separate the compounds with a gas chromatograph (GC) before they enter a mass spectrometer (GC-MS). Unfortunately, GCs are inherently slow – a typical separation of just one sample can easily take 30 minutes – so while GC-MS is fine for analysing a single sample or monitoring slowly changing situations, it cannot usually be regarded as a "real time" or on-line technique.

In order to overcome these problems a new technique called Proton Transfer Reaction Mass Spectrometry (PTR-MS) was developed about ten years ago in our laboratories at the Institut für Ionenphysik of the Leopold Franzens Universität Innsbruck (LFUI), allowing quantitative on-line monitoring of volatile organic compounds (VOC). The fundamental difference between a conventional MS and PTR-MS is the "*soft ionisation*" method used to ionise the organic molecules. PTR-MS uses chemical ionisation, in which the VOC molecules are reacted in a drift tube with charged ions, in most cases hydroxonium ions (H_3O^+) produced in an external glow discharge ion source. The great advantage of this method, besides immediately giving absolute concentrations, is that fragmentation of the molecules is very much reduced so the mass spectra produced are much easier to interpret and are more straightforward to quantify. This means that for many quantitative applications the preliminary GC separation becomes unnecessary.

This novel technique enables a variety of organic species in complex matrices to be monitored in real-time (within seconds), with detection limits as low as a few parts per trillion, volume (pptv). Thus, one molecule out of 200 billion "air" molecules can be detected virtually in real time, without any pre-concentration procedure. PTR-MS therefore makes it possible to visualise the change of complex matrices in real time, yielding a fingerprint "film" and not just a fingerprint "snapshot".

In 1998 a spin-off company called Ionicon Analytik GmbH (www.ptrms.com) was founded to provide this technique to a growing user community. Today more than 80 instruments are in use throughout the world, including by noted multinational companies and renowned research institutions in the fields of environmental sciences, food technology and medicine.

The intent to found and organise the 1st International PTR-MS Conference in January 2003 in Igls, Austria was to bring together active scientists and technologists involved in real-world mass spectrometric measurements of VOC from both academia and industry. After this conference, which was organised to honour the late Professor Werner Lindinger, the pioneer of PTR-MS who died in 2001 in a tragic accident, the idea arose to summarise the present status of this field in a special issue of the International Journal of Mass Spectrometry. We were invited to serve as guest editors of this special Volume 239, Issues 2-3, finally published on 15th December, 2004. The number, variety and quality of published papers are themselves a testimony to the rapidly evolving PTR-MS field.

The contributions to the 2nd International PTR-MS Conference held in January/February 2005 in Obergurgl, Austria range from fundamentals and instrumentation of PTR-MS to a number of applications across disciplines such as environmental, food, flavour and life sciences, and demonstrate even more clearly how fast this field is growing.

We are honoured by the part we could play in collecting, editing and publishing the special PTR-MS issue and this Conference Proceedings. We would like to thank Christine Oberhofer who contributed so much in organising this conference and Wolfgang Sailer who was responsible for the Conference web page (www.ptrms-conference.com). Special thanks go to Armin Wisthaler and Wolfgang Grabmer who supported the organisation and editing of the Conference Proceedings.

Armin Hansel
Tilmann Märk

January 2005, Innsbruck

www.ptrms.com

Ionicon Analytik GmbH
Technikerstrasse 21a
6020 Innsbruck
Austria (Europe)

Phone: +43 (0)512 50748 00
Fax: +43 (0)512 50798 18
E-Mail: info@ptrms.com
Web: **www.ptrms.com**

Proton-**T**ransfer-**R**eaction **M**ass-**S**pectrometry

Volatile **O**rganic **C**ompounds - Detector

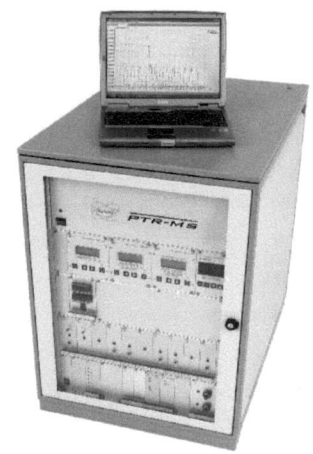

- **Mass range:**   1 – 512 amu
- **Measuring time:**   2 – 60.000 ms/amu
- **Response time:**   < 200 ms
- **Linearity range:**   10 ppt – 5 ppm
- **Inlet temperature:**   adjustable; 30 – 80 °C
- **Drift tube temperature:**   adjustable; 30 – 80 °C
- **Inlet flow:**   adjustable; 15 – 200 sccm
- **Weight:**   130 kg
- **Power consumption:**   115 V / 230 V – 700 W
- **Dimensions [cm]:**   78 x 86 x 55 (w x h x d)

compact
PTR-MS

Proton-**T**ransfer-**R**eaction **M**ass-**S**pectrometry

Volatile **O**rganic **C**ompounds - Detector

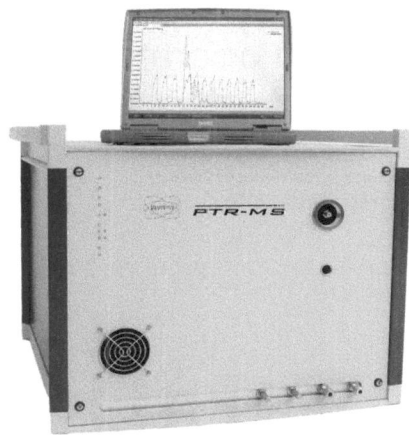

- **Detection limit:** 500 pptv
- **Mass range:** 1 – 300 amu
- **Measuring time:** 2 – 60.000 ms/amu
- **Response time:** < 200 ms
- **Weight:** approx. 55 kg
- **Power consumption:** approx. 300 W
- **Dimensions [cm]:** 55 x 44 x 60 (w x h x d)
- **Operation:** remote control via PC (Ethernet)

A Compact Proton-Transfer-Reaction Mass-Spectrometer (PTR-MS) for use in Food Research, Medicine, and Environmental Trace Gas Monitoring.

The real-time method uses H_3O^+ primary ions, which perform mostly non-dissociative proton transfer reactions with Volatile Organic Compounds (VOCs) present in air. Organic trace gases such as carbonyls, alcohols, aldehydes, BTX-compounds and many others are monitored within seconds with a detection limit of 500 pptv.

Contents

1. Opening Lecture

2. Invited Lectures

2.1. Environmental Science and Technology

2.2 Food Technology

2.3 Medical Applications

3. Contributed Papers (Posters)

1. Opening Lecture

An Anecdotal History of Ion-Molecule Reaction Flow Tube Studies from Genesis to PTR-MS

Eldon Ferguson

NOAA-CMDL, Boulder, Colorado

The Flowing Afterglow Ion Molecule reaction technique was developed in Boulder in 1964 for the purpose of elucidating the ion chemistry of the earth's ionosphere. This goal was accomplished as well as the determination of the positive and negative ion chemistry of the stratosphere. The FA made major contributions to ion thermochemistry, the understanding of energy transfer processes, and reaction mechanisms. Ultimately it led to the valuable analytical trace gas measurement technique PTR-MS.

Some examples of the analytical utility applied to in-situ atmospheric trace gas detection will be described. The origin of proton hydrates in the lower ionosphere was discovered and later applied to the discovery and measurement of acetonitile in the stratosphere by PTR. Other PTR type studies illuminated the chemical fate of sodium derived from atmospheric meteor ablation. Other CIMS type studies led to quantification of gas phase sulfuric acid in the stratosphere and nitric oxide in the lower ionosphere. Si^+ reactions with H_2O were utilized to make the highest altitude measurements (~90km) of H_2O.

2. Invited Lectures

2.1 Environmental Science and Technology

VOC flux measurements with PTR-MS: Possibilities and limitations

Albrecht Neftel, Aurelia Brunner and Christoph Spirig

Air Pollution/Climate Group, Agroscope FAL, Reckenholzstrasse 191, CH-8046 Zürich

VOCs play a major role in atmospheric chemistry. They are precursors of ozone and of a large fraction of secondary aerosols. Prognostic Chemistry Transport Models (CTMs) from the local to the global scale are the most important tool to evaluate optimal reduction strategies and possible feedback mechanisms between climate changes and changes on different ecosystem levels. Key information in this context is the determination of exchange fluxes of the large variety of VOCs at the biosphere-atmosphere interface. CTMs need reliable estimates of emissions and deposition of them. Measurements and estimates of these quantities represent an important issue in all environmental research clusters over the last 50 years. A major challenge is the determination of fluxes over different ecosystems due to the large spatial inhomogeneities and temporal dynamics. The most advanced method to determine exchange fluxes is the eddy correlation technique, which directly determines the vertical flux of a trace component by measuring the covariance between fluctuations in vertical wind velocity and the mixing ratio of the trace components. The analytical eye of the needle for a multicomponent flux determination is the availability of an appropriate chemical sensor capable of making fast and sensitive measurements of a large variety of compounds. The PTR-MS technique now allows fast and sensitive VOCs measurements satisfying the requirements for doing EC flux measurements of the relevant compounds.

An impressive amount of publications has appeared in the scientific literature demonstrating the potential of PTR-MS for flux measurements. We report measurements over grassland and forests. Limitations in the interpretation of the flux measurements will be discussed in light of analytical, meteorological and site-specific characteristics.

Biosphere-Atmosphere Exchange Investigated Using PTR-MS

Thomas Karl, Alex Guenther

National Center for Atmospheric Research, P.O. Box 3000, Boulder, 80305, USA,
tomkarl@ucar.edu

ABSTRACT

Proton-transfer-mass spectrometry has proven to be a valuable tool for investigating surface exchange by means of the eddy covariance and the gradient method. Results from two recent field campaigns – the CELTIC (Chemical Emission, Loss, Transformation and Interactions within Canopies) and CAPOS (Chemistry And Production Of Smoke in Brazil) studies - will be presented and put into the context of upscaling primary biogenic emissions and investigating the impact on atmospheric chemistry in the boundary layer.

1. Introduction

Organic molecules are the signature compounds of the biosphere. They exist in various phases including volatile organic compounds (VOC) that are emitted into the air where they can have a substantial impact on the chemistry of the atmosphere. Terpenoid and oxygenated compounds are among the most abundant biogenic VOCs and comprise about 80-90% of the total emission strength estimated to be on the order of 1000 Tg/y. [Guenther et al., 1995]. The main challenge in upscaling biogenic VOC emissions is the heterogeneity of these fluxes due to plant diversity especially in tropical ecosystems. The PTR-MS technology has allowed expanding our knowledge on the processes governing the exchange of VOCs putting more rigorous constraints on global budgets in recent years. Despite the progress there are still challenges ahead, especially when synthesizing observations into atmospheric models. The IPCC [2001] report notes that the use of larger isoprene fluxes in the IPCC CTMs (chemistry transport models) resulted in poor agreement with observations of CO and isoprene concentration. This lead the IPCC [2001] report to conclude that current bottom up approaches overestimate emissions into the atmosphere by not accounting for the fraction that does not escape from the vegetation canopy. Latest observations using the fast measurement capabilities of the PTRMS system and it's ability to measure isoprene plus it's first generation oxidation products MVK+MAC suggested a canopy reduction factor of at most 10%. This is substantially lower than suggested in the IPCC report. These observations also highlighted other uncertainties such as our inability to model dry deposition of many VOCs.

2. Results

2.1 Emission Modeling

Figure 1 shows the source/sink distribution calculated from an inverse lagrangian transport model using an incanopy profiling system [Karl et al., 2004]. The shaded areas depict the tree composition at this particular site indicating that monoterpene emissions are dominantly emitted from pine trees, whereas isoprene is emitted from sweetgum trees in the understory. Methanol emissions correlate with the pine tree leaf area index. These results illustrate implications for emission modeling in the eastern US as it shows that sweetgum trees in the understory can emit a major fraction of reactive species (isoprene). Landcover data are based on remote sensing, which typically do not classify understory species. In this particular case this would result in a substantially underestimated airmass reactivity if not corrected. Emission profiles such as the one shown in figure 1 also help to improve upscaling efforts in tropical ecosystems where leaves are notoriously hard to access and the high species diversity (eg. ~100-200 species / ha) prevents screening of individual trees in a representative manner.

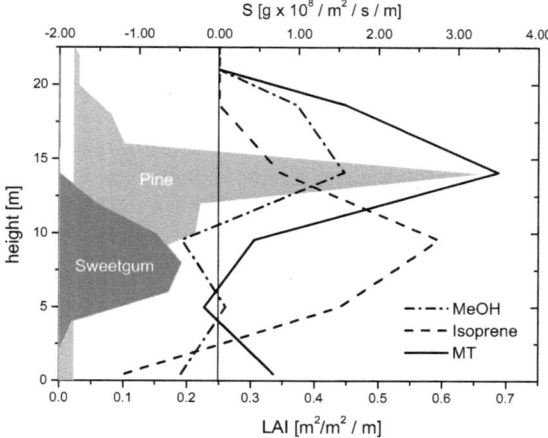

Figure 1: Source/Sink distribution of Methanol (MeOH), Isoprene and Monoterpenes (MT) (top axis) plotted together with the Differential Leaf Area index (LAI) (bottom axis) at Duke Forest

2.2 Implications for Atmospheric Chemistry and Transport

The inability to quantitatively predict the atmospheric fate of isoprene in the tropics by most if not all global atmospheric chemistry and transport models based on current best estimates of isoprene emissions highlights the need for more measurements. Results from airborne and ground based flux measurements in the Amazon during the CAPOS 2004 study point towards several gaps in our understanding when quantifying sources and sinks of VOCs in the boundary layer:

(1) The deposition of moderately soluble species is higher than predicted using conventional dry deposition schemes.
(2) The effect of clouds does not only influence primary surface emissions, but also modifies turbulent exchange in the mixed layer which might currently be overpredicted by transport schemes
(3) A shallow cumulus cloud cover seems to enhance the capacity to oxidize various VOCs in the cloud layer.

Acknowledgements: We thank Bob Yokelson from the University of Montana, Paulo Artaxo from the University in Sao Paulo, Mark Potosnak from the Desert Research Institute and the CELTIC science team (http://www.acd.ucar.edu/CELTIC/) for their support and collaboration. The National Center for Atmospheric Research is sponsored by the National Science Foundation.

References:
Guenther, A., C.N. Hewitt, D. Erickson, R. Fall, C. Geron, T. Graedel, P. Harley, L. Klinger, M. Lerdau, W.A. Mckay, T. Pierce, B. Scholes, R. Steinbrecher, R. Tallamraju, J. Taylor, and

P. Zimmerman, A Global-Model of Natural Volatile Organic-Compound Emissions, Journal of Geophysical Research-Atmospheres, 100 (D5), 8873-8892, 1995.

IPCC, The Scientific Basis, Chapter 4.2.3.2; Atmospheric Chemistry and Greenhouse Gases, Volatile organic compounds, http://www.ipcc.ch/, 2001.

Karl, T., M. Potosnak, A. Guenther, D. Clark, J. Walker, J.D Herrick, and C. Geron, Exchange Processes of Volatile Organic Compounds above a Tropical Rainforest – Implications for Modeling Tropospheric Chemistry above Dense Vegetation, J. Geophys. Res., Vol. 109, No. D18, D18306, 10.1029/2004JD004738, 2004.

Terpenes and their oxidation products in a pine forest: insights from novel PTR-MS measurements

Rupert Holzinger[1], Anita Lee[1], Gunnar Schade[2], Allen H. Goldstein[1]

[1] University of California at Berkeley, USA (holzing@nature.berkeley.edu)

[2] Now at Institute for Environmental Physics, University of Bremen, Germany

ABSTRACT

Trace gas measurements have been made at the Blodgett forest site since 1997. Methanol, acetone, methylbutenol (MBO), isoprene, and monoterpenes (α-pinene, β-pinene, Δ-3-carene, limonene, and others) have been identified as major biogenic emissions from the local vegetation (Lamanna and Goldstein 1999). Eddy covariance measurements of ozone flux revealed a large deposition of ozone into the ecosystem. Detailed analysis and modeling efforts by (Kurpius and Goldstein 2003) showed that during the summer less than 50% of the deposition could be explained by dry deposition and stomatal uptake. The residual flux (i.e. the total flux minus the flux that could be attributed to stomatal uptake and dry deposition) was hypothesized to be due to chemical reactions within the forest canopy. This chemical ozone flux scaled with temperature in the same way (but opposite sign) as monoterpenes, so reaction with short lived terpenoid compounds were suspected to account for the observed chemical ozone loss. However, above-canopy fluxes of speciated monoterpenes above the canopy, measured with an automated GC-FID-REA system, were an order of magnitude lower than the chemical ozone flux, thereby suggesting that the majority of local terpene emission is lost within the forest canopy. Subsequently, a forest thinning experiment was performed which simultaneously and dramatically enhanced both monoterpene emissions and ozone deposition, providing confirmation that ozone deposition was dominated by reactions with biogenic VOCs (Goldstein, McKay et al. 2004). In search of these missing emissions and their oxidation products, PTR-MS measurements were started at the Blodgett forest site.

1. Introduction

The Blodgett forest site on the western slope of the Sierra Nevada Mountains of California (38.90O N, 120.63O W, 1315m elevation) is located 75 km downwind (northeast) of Sacramento where it receives anthropogenically impacted air masses rising from the valley below during the day (Bauer, Hultman et al. 2000). At night the wind usually shifts towards the west and air masses descending from the sparsely populated Sierra Nevada are advected to the site. The plantation is owned and operated by Sierra Pacific Industries (SPI) and was planted with Pinus ponderosa L. in 1990, interspersed with a few individuals of Douglas fir, white fir, California black oak, and incense cedar. Average tree height was 4.8 (median) in 2003 and the canopy height was 6.4m, a height not exceeded by 80% of the trees. The understory was composed primarily of manzanita (Arctostaphylos spp.) and whitethorn (Ceonothus cordulatus) shrubs. A sketch of the experimental setup is given in Figure 1. The air could be sampled through 6 individual gas inlets. Inlet A was used to sample air from 12.5m for eddy-covariance flux measurements of total monoterpenes. This inlet was located at the top of the tower, adjacent to the sonic anemometer. Air was pulled at 10 L min^{-1} through a 2 μm Teflon particulate filter, and brought down, using 6mm ID Teflon tubing, to the instruments in a temperature controlled container. Inlets B-F were used to sample vertical gradients at height-levels within (1.1, 3.1, 4.9m) and above (8.75, 12.5m) the canopy. The 5 gradient inlets each consisted of 30m PFA tubing (ID ~4mm) protected by a Teflon filter (PFA holder, PTFE membrane, pore size 2µm); a sample flow of 1 liter per minute was maintained at all times through each sample tube.

Figure 1. Experimental Setup: high flow inlet A was used for flux measurements; VOC-gradients were measured by sampling from 5 individual inlets (B-F) positioned at 1.1, 3.1, 4.9, 8.75, and 12.5m above ground. Canopy height was 6.4 m.

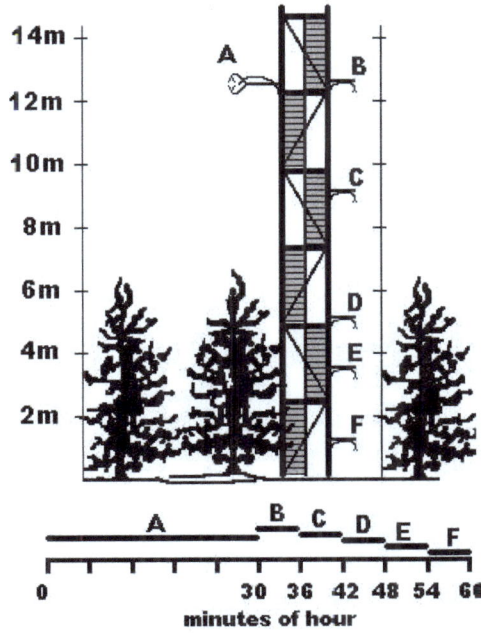

2. Results

The main focus for the first PTR-MS measurements at this site in 2002 was eddy-covariance fluxes of total monoterpenes above the canopy (Lee, Schade et al. 2004). Many monoterpenes have been identified in forest emissions using gas chromatography (GC). However, it has been impossible to determine whether all monoterpenes are appropriately measured using GC techniques. Speciated flux measurements of eight monoterpene species were made using a dual channel GC-FID combined with the relaxed eddy accumulation (REA) technique. The PTR-MS-EC and GC-FID-REA systems sampled from the same gas inlet system (including the sonic anemometer), and shared the same monoterpene gas-standards. So, relative differences could be monitored with high accuracy.

The diurnal cycles of total (PTR-MS) and the sum of speciated (GC-FID) monoterpenes over a four-day period in early August agreed well (Figure 2a), with higher mixing ratios measured by the PTR-MS. Mixing ratios of total monoterpenes averaged $30 \pm 2.3\%$ larger than mixing ratios of the sum of eight monoterpene species. Total monoterpene mixing ratios were $35 \pm 3.5\%$ larger at night and $19 \pm 2.5\%$ larger during the day (0700 – 1900 PST) than the sum of speciated monoterpenes, suggesting the absence of daytime photochemistry contributes to the detection of additional compounds that are typically lost during the day. The time series of the percent difference between the mixing ratios measured by PTR-MS and GC-FID, plotted with O3 mixing ratio, show that periods when the PTR-MS measures significantly more monoterpene than the GC-FID coincide with periods of low O3 mixing ratio (Figure 2b). Six to ten additional peaks were resolved by the GC-FID; four of these peaks correlated with β-pinene and accounted for the daytime discrepancy between PTR-MS and GC-FID mixing ratios. Despite the detection of eight identified and four unidentified potential monoterpenes the PTR-MS still measured ~20% more monoterpenes at night; thus the additional compounds detected by PTR-MS during the night must undergo photochemical reactions before detection in daytime. This is consistent with the daytime observations of oxidation products in the forest canopy that have been made during summer 2003 (Holzinger, Lee et al. 2004).

Figure 3 a-d shows vertical gradients and diurnal cycles for 4 compounds. To create these images we interpolated median values calculated from the 47 days for each compound, hour and height-level.

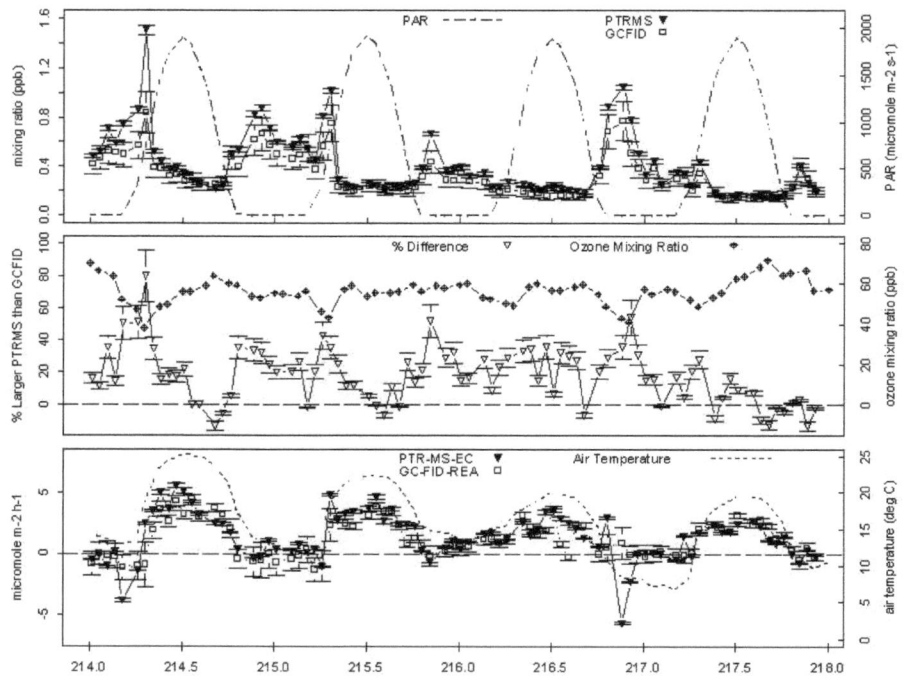

Figure 2: Total monoterpene mixing ratios (a) and fluxes (c) measured by PTR-MS are greater than the sum of eight monoterpene species measured by GC-FID. Diurnal cycle of the percent difference between PTR-MS and GC-FID mixing ratios, plotted with O3 mixing ratio, show that the PTR-MS measures more monoterpenes when O3 mixing ratios are lower (b).

Higher concentrations in the canopy provide a clear signature showing that terpenes (Figure 2a) are emitted from the ecosystem. Temperature is the main driver for monoterpene emissions. The short lifetimes of terpenes (minutes to hours) result in low concentrations during the day as compared to nighttime when both oxidation and vertical mixing are slower. MBO (Figure 2b) is another biogenically emitted compound whose emission has been previously reported from this site. In contrast to monoterpenes, the emission of MBO also requires light; therefore both the mixing ratios and gradients reach minimum values at night when no emission occurs. Nopinone (Figure 2c) is known to be produced from the oxidation of β-pinene by OH. Considering only gas phase chemistry, increasing mixing ratios would be expected during the day since the OH-lifetime of nopinone is about 5 times that of β-pinene and relatively constant β-pinene concentrations were usually observed from morning through afternoon, however, the contrary is observed. The decrease of nopinone from morning to afternoon is anti-correlated with increases in aerosol concentration over the same period. We conclude that the decrease of nopinone through the day provides evidence of partitioning into the particle phase. The signal detected at mass 113 (OX02, Figure 2d) is representative of a whole class of compounds. These compounds behave similarly, and have so far only been identified by their mass to charge ratios (see Table 1). Their diurnal and vertical profiles share the following characteristics: (i) higher concentration above than within the canopy; (ii) significantly lower concentration at both the top and the bottom levels; and (iii) nighttime concentrations significantly lower than daytime concentrations. Because the concentration maximum is spatially separated from any primary emission source, these compounds must be oxidation products of primary biogenic compounds whose emissions are highest during daytime.

Using a simplified surface renewal model we estimated the emission of biogenic precursors compounds to be in the range of 13–66 $\mu mol\ m^{-2}\ h^{-1}$. That is 6–30 times the emissions of total monoterpenes observed above the forest canopy on a molar basis.

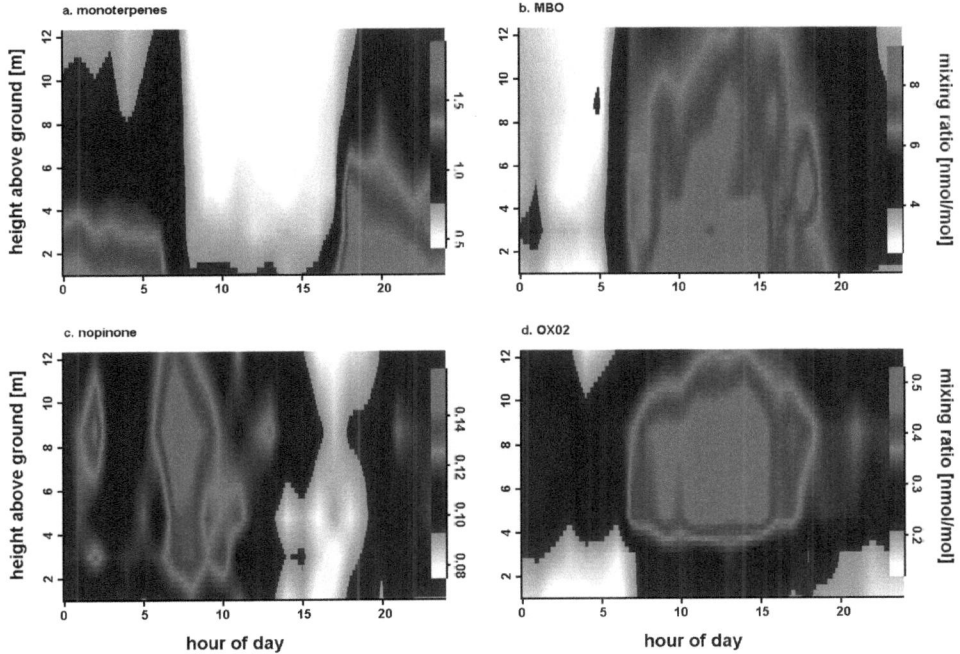

Figure 3. Vertical gradients for individual compounds. Measurements over 47 days were averaged to produce profiles representative of summer 2003. Monoterpenes (a) and MBO (b) are examples of primary biogenic emissions. Decreasing mixing ratios during the day are indicative of heterogeneous chemistry of nopinone (c). Highest concentrations above the canopy identify OX02 (d) as an oxidation product with a strong local source.

Bauer, M. R., N. E. Hultman, J. A. Panek and A. H. Goldstein (2000). "Ozone deposition to a ponderosa pine plantation in the Sierra Nevada Mountains (CA): A comparison of two different climatic years." Journal of Geophysical Research-Atmospheres **105**(D17): 22123-22136.

Goldstein, A. H., M. McKay, M. R. Kurpius, G. W. Schade, A. Lee, R. Holzinger and R. Rasmussen (2004). "Forest thinning experiment confirms ozone deposition to forest canopies is dominated by reaction with biogenic VOCs." Geophysical Research Letters **31**(doi:10.1029/2004GL021259): in press.

Holzinger, R., A. Lee, K. T. Paw U and A. H. Goldstein (2004). "Observations of oxidation products above a forest imply biogenic emissions of very reactive compounds." Atmos. Chem. Phys. Discuss. **4**: 5345-5365.

Kurpius, M. R. and A. H. Goldstein (2003). "Gas-phase chemistry dominates O-3 loss to a forest, implying a source of aerosols and hydroxyl radicals to the atmosphere." Geophysical Research Letters **30**(7): art. no.-1371.

Lamanna, M. S. and A. H. Goldstein (1999). "In situ measurements of C-2-C-10 volatile organic compounds above a Sierra Nevada ponderosa pine plantation." Journal of Geophysical Research-Atmospheres **104**(D17): 21247-21262.

Lee, A., G. W. Schade, R. Holzinger and A. H. Goldstein (2004). "A comparison of new measurements of total monoterpene flux with improved measurements of speciated monoterpene flux." Atmos. Chem. Phys. Discuss.(submitted).

Bauer, M. R., N. E. Hultman, et al. (2000). "Ozone deposition to a ponderosa pine plantation in the Sierra Nevada Mountains (CA): A comparison of two different climatic years." Journal of Geophysical Research-Atmospheres **105**(D17): 22123-22136.

Goldstein, A. H., M. McKay, et al. (2004). "Forest thinning experiment confirms ozone deposition to forest canopies is dominated by reaction with biogenic VOCs." Geophysical Research Letters **31**(doi:10.1029/2004GL021259): in press.

Holzinger, R., A. Lee, et al. (2004). "Observations of oxidation products above a forest imply biogenic emissions of very reactive compounds." Atmos. Chem. Phys. Discuss. **4**: 5345-5365.

Kurpius, M. R. and A. H. Goldstein (2003). "Gas-phase chemistry dominates O-3 loss to a forest, implying a source of aerosols and hydroxyl radicals to the atmosphere." Geophysical Research Letters **30**(7): art. no.-1371.

Lamanna, M. S. and A. H. Goldstein (1999). "In situ measurements of C-2-C-10 volatile organic compounds above a Sierra Nevada ponderosa pine plantation." Journal of Geophysical Research-Atmospheres **104**(D17): 21247-21262.

Lee, A., G. W. Schade, et al. (2004). "A comparison of new measurements of total monoterpene flux with improved measurements of speciated monoterpene flux." Atmos. Chem. Phys. Discuss.(submitted).

Alternative Carbon Sources for Leaf Isoprene Formation

M. Graus[1], J.-P. Schnitzler[2], J. Kreuzwieser[3], U. Heizmann[3], H. Rennenberg[3], A. Wisthaler[1], and A. Hansel[1]

[1]*Institut für Ionenphysik, Leopold Franzens Universität Innsbruck, Innsbruck, Austria (martin.graus@uibk.ac.at)*

[2]*Forschungszentrum Karlsruhe, Institut für Meteorologie und Klimaforschung Atmosphärische Umweltforschung, Garmisch-Partenkirchen, Germany*

[3]*Institut für Forstbotanik und Baumphysiologie, Albert-Ludwig-Universität Freiburg, Freiburg, Germany*

ABSTRACT

Isoprene formation in leaves is closely linked to photosynthesis; however, the carbon incorporated does not originate entirely from recently assimilated CO_2. Isotopic labelling (^{13}C), in feeding and fumigation experiments, proofs the existence of additional carbon sources. $^{13}CO_2$ fumigation in our studies with young poplar (*Populus* × *canescens*) trees showed only ~75% labelling in the isoprene. The consequence of feeding ^{13}C glucose via xylem was ~10% labelling, fill-up of internal carbon pools (like starch) resulted in ~30% ^{13}C incorporation. PTR-MS technique turned out to be a very efficient tool to quantify leaf isoprene emissions and to follow online the rapid incorporation of carbon from different, isotope-labelled sources and the fast wash out of labelling.

1. Introduction

Many plant species, particularly trees, emit isoprene by photosynthesising leaves [1]. Global atmospheric flux is estimated to be 503 million tons of carbon per year and supplies ~44% to the total VOC emissions of biogenic origin [2]. Understanding biochemical and physiological mechanisms controlling the formation of isoprene in plants and its emission is of interest as isoprene chemistry can contribute substantially to ozone production [3]. It also has an influence on the tropospheric OH concentration and therewith indirect effect on the abundance of methane, an important greenhouse gas [4].

Light and temperature control isoprene emission, and simple emission models using these two parameters and a standard emission factor can predict almost all short-term variations of potential isoprene emissions [5]. Variations in diurnal isoprene fluxes that cannot be explained (e. g. under severe water stress [6] or midday depression of photosynthesis [7]), have led to the suggestion that alternative carbon sources may also contribute to the production of isoprene [8]. Alternative carbon can originate for instance form xylem-transported carbon compounds [9], breakdown of starch [10][11] or mitochondrial respiration [12].

Our studies aimed to quantify carbon sources, apart from photo-assimilates and xylem-derived sugars, that contributed to isoprene formation in poplar (Populus x canescens) leaves. For this purpose, leaf internal carbon pools were labelled by $^{13}CO_2$ via photosynthesis and the contribution of this carbon pool to isoprene formation was quantified with on-line proton-transfer-reaction mass spectrometry (PTR-MS). In addition to this, feeding experiments with [1,2-^{13}C]glucose and [3-^{13}C]glucose were performed to trace and quantify the transition of the ^{13}C label from cytosolic pyruvate / phospho*enol*pyruvate equivalents into the chloroplastidic DOXP-pathway.

2. Results and Discussion

2.1. Exposure to $^{13}CO_2$ causes fast but incomplete ^{13}C labelling of emitted isoprene

Isoprene is closely linked to photosynthesis. Dynamic use of assimilated carbon for isoprene biosynthesis in intact leaves can be demonstrated in $^{13}CO_2$ labelling experiments [13][14][15]. Our studies with poplar show that 20-30% of isoprene carbon atoms remained unlabelled when leaves were exposed to $^{13}CO_2$, even under non-stressed conditions with net assimilation rates in the same range as found in other experiments with poplar hybrids of the same age, from the same line [16]. Along with previous $^{13}CO_2$ fumigation experiments by Delwiche and Sharkey [14] and Karl et al. [15] the results of these studies support the view that readily available carbon pools, which are not directly linked to recently assimilated carbon, exist in leaves to enable formation of the direct isoprene precursor dimethylally diphosphate (DMADP).

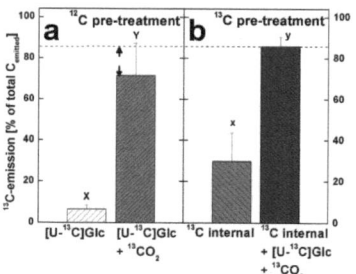

Figure 1. Portion of ^{13}C from total carbon emitted as isoprene as response to different labelling treatments.

2.2. Alternative Carbon Sources

Karl et al. [15], and Affek and Yakir [17] proposed that additional carbon for isoprene formation could be derived from chloroplastidic breakdown of starch, occurring simultaneously to starch biosynthesis, re-fixation of respiratory carbon, or influx of cytosolic precursors (pyruvate/PEP) into the chloroplast.

2.2.1. Xylem Transported Carbon

To investigate whether xylem transported carbohydrates account for isoprene biosynthesis, excised poplar leaves were fed 5mm ^{13}C-labelled glucose via the xylem. Within minutes, labelling appeared in the emitted isoprene and the portion of incorporated ^{13}C amounted to ~4-10% at stabilised conditions 30min after treatment. In addition to [U-^{13}C]Glc fed via the xylem, poplar leaves were exposed to $^{13}CO_2$. Still only 72% ± 10% of carbon emitted as isoprene was labelled some 30min after treatment change, indicating that other carbon sources must exist.

2.2.2. Leaf Internal Carbon Pools

Besides photo-assimilates and xylem-derived sugars for isoprene formation in poplar leaves, we quantified additional leaf internal carbon pools. To determine their contribution to isoprene formation, these leaf internal carbon pools were labelled with ^{13}C by keeping the trees in darkness for 48hrs to deplete leaf internal C pools (in particular starch). Individual leaves were subsequently exposed to 360ppm$_v$ $^{13}CO_2$ in cuvettes during a continuous irradiation period of three days.

Continuous irradiation was chosen to prevent starch synthesis degradation occurring in poplar leaves during the night. In these experiments starch concentration decreased to ~40% of the initial levels; thus, a remarkable unlabelled pool of starch remained and was still present during exposure to $^{13}CO_2$. Starch concentrations recovered to control levels during the light period

After this pre-treatment, emission measurements during exposure to 360ppm$_v$ $^{13}CO_2$ were started (Fig. 2). Total isoprene emission of poplar leaves ranged from 5-12 nmol m^{-2} s^{-1} (e.g. Fig. 2a). Isotope mass m74$^+$ dominated these emissions, amounting to ~4 nmol m^{-2} s^{-1} (Fig. 2a). Emissions of isoprene isotopes m73$^+$ (~3.5 nmol m^{-2} s^{-1}), m72$^+$ (~1.7 nmol m^{-2} s^{-1}), m71$^+$ (0.7 nmol m^{-2} s^{-1}), m70$^+$ (0.2 nmol m^{-2} s^{-1}) and m69$^+$ (0.1 nmol m^{-2} s^{-1}) were significantly lower. Switching from $^{13}CO_2$ to $^{12}CO_2$ exposure caused a drastic exchange of isotope distribution. The total isoprene emission rates (sum of all isoprene isotopes), however, did not change. During exposure to $^{12}CO_2$ the portion of ^{13}C of total carbon emitted as isoprene still amounted to ~30 ± 12% (see also Fig. 1b). This labelling of isoprene presumably derived from starch or other leaf internal carbon pools labelled during pre-treatment with $^{13}CO_2$. Such values agree with actual measurements of the natural abundance of carbon isotope composition ($\delta^{13}C$), which demonstrated that 9-28% of isoprene carbon was contributed from alternative, slow turnover, carbon source(s) [17] for three different isoprene-emitting species.

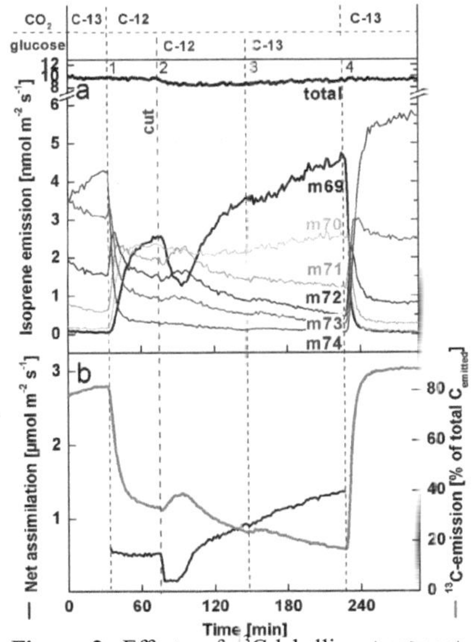

The nature of additional carbon source(s) for isoprene formation is still unclear. Use of starch for isoprene formation requires its partial breakdown to occur simultaneously to its synthesis, a feature which has been described for spinach chloroplasts [18] but has yet to be demonstrated for poplar. A more likely explanation - supported by observed rapid changes in the isoprene isotopes m71$^+$ and m72$^+$ when net assimilation is fluctuating (see for example after excising the intact leaf from the tree, Fig. 2, section 2) - is the assumption that cytosolic sugar and carbon compounds, and plastidic carbon compounds, which do not directly participate in photosynthetic carbon reduction, act as an alternative carbon source for isoprene formation.

Further tests were conducted to establish whether [U-^{13}C]Glc fed to leaves via petioles increases the rate of ^{13}C-labelling of isoprene in addition to its labelling by leaf internal ^{13}C labelled carbon pools. Additional feeding of [U-^{13}C]Glc counteracted the slow,

Figure 2. Effects of ^{13}C-labelling treatment (glucose feeding and CO_2 fumigation) on (a) the emission of total isoprene and its different isotopes and (b) the portion of ^{13}C from total carbon emitted as isoprene and net assimilation.

continuous washout of the ^{13}C label and led to a small, transient increase in emissions of double and treble labelled isotopes (Fig. 2a). The concurrent emission of the unlabeled isotope species (m69$^+$) dropped. The effect of additional ^{13}C via the xylem in the study with ^{13}CO$_2$ pre-treated plants was minimal compared to the exclusive feeding of [U-^{13}C]Glc shown in Fig. 1a. strongly indicating that the cytoplasmic pool of glycolytic intermediates was widely labelled with ^{13}C and that additional xylem-derived ^{13}C only slightly affected cytosolic ^{13}C sources that contributed to isoprene formation.

On completion of experiments shown in Fig. 2, ^{13}C supply from leaf internal carbon sources - xylem-transported [U-^{13}C]Glc, plus atmospheric ^{13}CO$_2$ - resulted in an overall ^{13}C labelling rate of isoprene molecules of 85% . This indicates that a complete labelling of the isoprene molecule was not obtained. The gap of ~15% unlabelled carbon in isoprene may be due to (a) the incomplete removal of unlabelled starch at the start of ^{13}CO$_2$ fumigation, (b) an incomplete exchange of carbon in pools with low turnover rates, or (c) the fact that mature poplar leaves fumigated with ^{13}CO$_2$ received unlabelled carbon compounds via xylem and phloem. Inconsistent with the finding of a ~30% use of alternative carbon sources for isoprene formation (Fig. 1b), a considerably lower ^{13}C amount was found in the isoprene emitted from leaves labelled by xylem-transported glucose plus atmospheric ^{13}CO$_2$ (Fig. 1a) than from leaves labelled by leaf internal, xylem-transported plus atmospheric ^{13}CO$_2$ (Fig. 1b) (see arrows in Fig. 1a); this is indirect evidence for a contribution of internal carbon sources.

2.3. Summary of Further Results of these Experiments (Details can be found in [9] and [19])

- Turnover in isoprene labelling after changing from ^{13}CO$_2$ exposure to ^{12}CO$_2$ and vice versa occurs on two timescales.
- Transiently decreased net assimilation causes enhanced incorporation of alternative carbon sources into emitted isoprene.
- Feeding glucose with ^{13}C atoms on diverse positions in the sugar molecule [U-^{13}C]Glc, [1,2-^{13}C]Glc, [3-^{13}C]Glc) indicate a direct incorporation of a C_2 fragment into isoprene.

3. Conclusion

Besides photosynthetically fixed CO_2 in poplar leaves, other carbon sources are used for isoprene formation contributing to about 20-30% of the total carbon emitted as isoprene. These alternative sources become even more important under limited photo-assimilation of CO_2. The nature of these carbon sources is quite heterogeneous; about one third seems to be derived from xylem transported carbon compounds such as carbohydrates. The rest could be provided either from starch degradation or from other carbon containing compounds present in the leaves. Future studies should follow two strategies: First, the interactions between chloroplasts and the cytosol should be investigated in order to understand which carbon compounds are shifted between the two compartments and how these processes are regulated. Secondly, the nature of carbon sources other than CO_2 and xylem transported glucose used for isoprene formation should be identified.

4. References

1. **Kesselmeier J, Staudt M** (1999) Biogenic volatile organic compounds (VOC): an overview on emission, physiology and ecology. J Atmos Chem **33**: 23-88
2. **Guenther AB, Hewitt CN, Erickson D, Fall R, Geron C, Graedel T, Harley P, Klinger L, Lerdau M, McKay WA, Pierce T, Scholes B, Steinbrecher R, Tallamraju R, Taylor J, Zimmerman P** (1995) A global model of natural volatile organic compound emissions. J Geophys Res **100**: 8873-8892
3. **Biesenthal TA, Wu Q, Shepson PB, Wiebe HA, Anlauf KG, MacKay GI** (1997) A study of relationships between isoprene, its oxidation products, and ozone, in the lower Fraser valley, BC. Atmos Environ **31**: 2049-2058
4. **Chameides WL, Lindsay RW, Richardson J, Kiang CS** (1988) The role of biogenic hydrocarbons in urban photochemical smog: Atlanta as a case study. Science **241**: 1473-1475
5. **Guenther AB, Zimmerman PR, Harley PC, Monson RK, and Fall R** (1993) Isoprene and monoterpene emission rate variability: Model evaluations and sensitivity analyses. J Geophys Res **98**: 12,609–12,618.
6. **Brüggemann N, Schnitzler J-P** (2002b) Comparison of isoprene emission, intercellular isoprene concentration and photosynthetic performance in water-limited oak (*Quercus pubescens* Willd. and *Quercus robur* L.) saplings. Plant Biol **4**: 456-463
7. **Zimmer W, Brüggemann N, Emeis S, Giersch C, Lehning A, Steinbrecher R, Schnitzler J-P** (2000) Process-based modeling of isoprene emission by oaks. Plant Cell Environ **23**: 585-597
8. **Sharkey TD, Yeh S.** (2001) Isoprene emission from plants. Annu Rev Plant Phys Plant Mol Biol **52**: 407-436
9. **Kreuzwieser J, Graus M, Wisthaler A, Hansel A, Rennenberg H, Schnitzler J-P** (2002) Xylem-transported glucose as additional carbon source for leaf isoprene formation in *Quercus robur*. New Phytol **156**: 171-178
10. **Monson RK, Harley PC, Litvak ME, Wildermuth M, Guenther AB, Zimmerman PR, Fall R** (1994) Environmental and developmental controls over the seasonal pattern of isoprene emission from aspen leaves. Oecologia **99**: 260-270
11. **Kelly GJ, Latzko E** (1995) Photosynthesis. In: Behnke HD, Lüttge U, Esser K, Kadereit JW, Runge M, eds. Prog Bot **56**: 134-163
12. **Anderson MD, Che P, Song J, Nikolau BJ, Syrkin-Wurtele E** (1998) 3-Methylcroronyl coenzyme A carboxylase is a component of the mitochondrial leucine catabolic pathway in plants. Plant Physiol **118**: 1127-1138
13. **Sanadze GA, Dzhaiani GI, Tevzadaze IM** (1972) Incorporation into the isoprene moelcule of 13CO2 assimilated during photosynthesis. Sov Plant Physiol **19**: 17-20
14. **Delwiche C, Sharkey TD** (1993) Rapid appearance of ^{13}C in biogenic isoprene when $^{13}CO_2$ is fed to intact leaves. Plant Cell Environ **16**: 587-591
15. **Karl T, Fall R, Rosenstiel TN, Prazeller P, Larsen B, Seufert G, Lindinger W** (2002a) On-line analysis of the $^{13}CO_2$ labelinglabelling of leaf isoprene suggests multiple subcellular origins of isoprene precursors. Planta **215**: 894-**905**
16. **Kreuzwieser J, Kühnemann F, Martis A, Rennenberg H, Urban W** (2000) Diurnal pattern of acetaldehyde emission by flooded poplar trees. Physiol Plant **108**: 79-86
17. **Affek HP, Yakir D** (2003) Natural abundance carbon isotope composition of isoprene reflects incomplete coupling between isoprene synthesis and photosynthetic carbon flow. Plant Physiol **131**: 1727-1736
18. **Stitt M, Heldt HW** (1981) Simultaneous synthesis and degradation of starch in spinach chloroplasts in the light. Biochim Biophys Acta **638**: 1-11
19. **Schnitzler J-P, Graus M, Kreuzwieser J, Heizmann U, Rennenberg H, Wisthaler A, and Hansel A** (2004) Contribution of Different Carbon Sources to Isoprene Biosynthesis in Poplar Leaves. Plant Physiol **135**: 152-160

A precise measurement system for plant's VOC uptake using proton transfer reaction mass spectrometry

Akira Tani[1], Nick Hewitt[2]

[1]School of High-Technology for Human Welfare, Tokai University, 317 Numazu, Shizuoka 410-0395, Japan. atani@wing.ncc.u-tokai.ac.jp.

[2]Institute of Environmental and Natural Sciences, Lancaster University, Lancaster LA1 4YQ, U.K.

ABSTRACT

In order to evaluate the contribution of higher plant to volatile organic compound (VOC) removal from atmosphere, a measurement system consisting of proton transfer reaction mass spectrometer (PTR-MS), CO_2 analyzer, diffusion devise and leaf enclose bag was established. Plant's VOC uptake, photosynthetic and transpiration rates could be simultaneously determined with this system. Compared with a measurement system based on GC, the system using PTR-MS can decrease uncertainties in determining plant's VOC uptake rate. The dynamic change in plant's VOC uptake rate could be also monitored with the measurement system. All species of ketones treated here, except for acetone, were uptaked by plants. The real-time monitoring using PTR-MS in this system might be a strong tool for measuring various VOC uptakes by plants.

1. Introduction

The contribution of vegetation to VOC removal from atmosphere has been of a great concern and the air to leaf transfer of VOCs is believed to be important as well as the VOC uptake by plant roots. Phenol is absorbed via stomata and transformed in leaf (Kondo et al., 1996). Formaldehyde is also absorbed and immediately metabolized (Kondo et al., 1995). Aromatic compounds such as benzene and toluene are accumulated in cuticular and resin (Keymeulen et al., 2001) and transformed inside leaf (Ugrekhelidze et al., 1997), although the uptake rates were not determined. Except for these compounds, however, little study has been conducted to know the plant ability to remove toxic VOCs.

Gas chromatography has been widely used to determine VOC concentrations. The uncertainty in sampling, injecting and analysing low concentrations (<1 ppmv) of VOCs is totally 3 to 10 % and this relatively large error has been a constraint to measure low VOC uptake rates of plant leaves. A recently developed instrument, proton transfer reaction mass spectrometry (PTR-MS), however, can measure VOC concentration directly and immediately, allowing us to get information about plant response to unstable and variable environments. The detection limit and response of the instrument is reported to ~10 pptv and within 1s, respectively (Hansel et al., 1998). No study has been conducted to determine VOC uptake rates by plants using PTR-MS.

In the present study, we established a VOC uptake measurement system using PTR-MS in order to obtain precise values of plant's VOC uptake rates. An overall Error in determining the VOC uptake rates was calculated and compared between PTR-MS and GC systems to know how precise the PTR-MS system is..

2. Experimental methods

VOC uptake measurement system using PTR-MS

VOC uptake measurement system consists of three parts: VOC-air preparation, fumigation and measurement parts (Fig. 1). A leaf of golden pothos (*Epipremnum aureum*) was enclosed with a transparent PFA bag (10~40 L) which had an inlet and outlet ports. The VOC air was introduced into the bag at a flow rate of 1.3 L/min. To measure both VOC and water vapour concentrations with a single PTR-MS instrument, the inlet and outlet lines were connected to PTR-MS via a three port solenoid valve which automatically changed lines every 5 minutes.

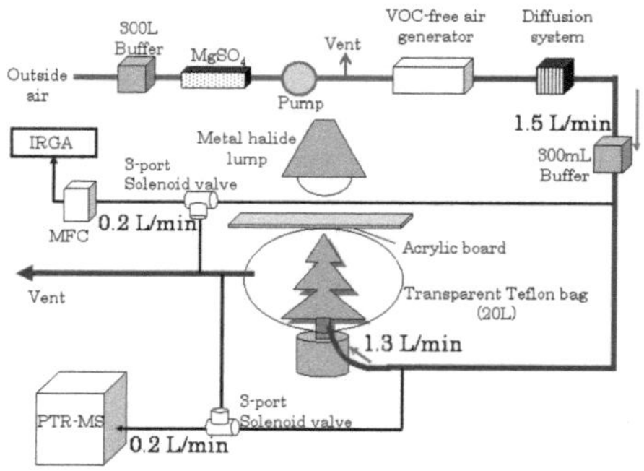

Fig. 1 Schematic diagram of a measurement system for plant's volatile organic compound uptake using proton transfer reaction mass spectrometry

PTR-MS operation

The drift tube E/N (where E is the electric field strength and N the buffer gas number density in the drift tube) was kept 125 Td by setting drift tube voltage, temperature and pressure at 550V, 40°C and 1.95 mbar, respectively. Since water vapour pressure was found to be correlated with the signal of mass 37 ($H_3O^+H_2O$) at an E/N of 125 Td, the ion was always monitored to measure water vapour pressure in the sample air and to determine plant transpiration rate.

Humidity correction

In this experiment, the outlet air contained a larger amount of water vapour resulting from plant transpiration and, therefore, humidity correction had to be made to determine VOC concentrations from PTR-MS signals. The dew point generator installed on the bypass line of the diffusion system was used to vary water vapour pressure in the air introduced to PTR-MS. The relationship between mass 19/37 ion ratio and VOC ion signal was obtained at various water vapour concentration.

Determination of VOC uptake rate and the corresponding error

Plant's VOC uptake rate A (mol m^{-2}s^{-1}) is determined by mass balance in the enclosure bag and calculated from

$$A = \frac{\text{Mass}_{in} - \text{Mass}_{out}}{A}, \tag{1}$$

where A is leaf area (m^2). Mass$_{in}$ and Mass$_{out}$ are VOC influx and efflux, respectively, per second. (mol s^{-1}) and determined from

$$\text{Mass}_{in} = V_{in} \times C^c_{in}, \tag{2A}$$

$$\text{Mass}_{out} = V_{out} \times C^c_{out}, \tag{2B}$$

where V_{in} and V_{out} are flow rates of the inlet and outlet air, respectively. (mol s^{-1}) and C^c_{in} and C^c_{out} are VOC concentrations of the inlet and outlet, respectively, corrected for humidity (mol mol^{-1}). C^c_{in} and C^c_{out} are determined from

$$C^c_{in} = \frac{C^r_{in}}{f^{VOC}_{in}} \tag{3A}$$

and

$$C^c_{out} = \frac{C^r_{out}}{f^{VOC}_{out}}, \tag{3B}$$

where C^r_{in}, and C^r_{out} are VOC concentrations of inlet and outlet, respectively, measured with PTR-MS. (mol mol^{-1}) and f^{voc}_{in} and f^{voc}_{out} are humidity correction terms for C^r_{in} and C^r_{out}, respectively (dimensionless).

Here VOC concentration, as well as water vapour, was measured every other 5 min, not simultaneously, for the inlet and outlet air streams. When calculating the VOC uptake and transpiration rates, the inlet values C^r_{in} and W^r_{in} were obtained as averages of the two data sets measured just before and after individual outlet data sets.

3. Results and discussion

Effect of water vapour on the measured concentration

Fig. 2 shows the relationship between mass ratio of 37 to 19 and concentrations of mass 57 and 101 determined with PTR-MS at different water vapour pressure. The relationship was determined for different MIBK concentrations (75-750 ppbv). Although the calculated concentration of mass 57, which is a fragment ion of MIBK, decreased with an increase in the mass 37/19 ratio, the calculated concentration of mass 101, which is a protonated molecular ion of MIBK, increased. This result suggests that correction must be made for water vapour in sample gas. The equations shown in Fig. 2 were used as f^{voc}_{in} and f^{voc}_{out} in Eqn (3A) and (3B).

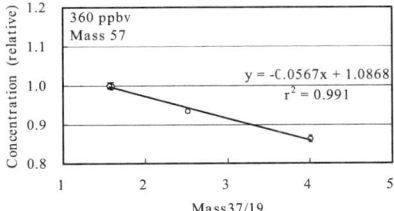

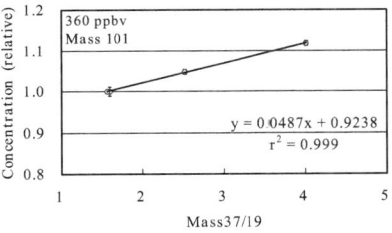

Fig. 2 Relationship between mass ratio of 37 to 19 and concentrations of mass 57 and 101 determined with PTR-MS at different water vapour concentration

Corrected concentration and error analysis

Fig. 3 shows changes in MIBK and water vapour concentrations of the inlet and outlet air streams. The plant leaf was exposed to four different concentrations of MIBK. The MIBK concentration of the inlet and outlet was very stable except for both the first and end data, where a series of mass measurement within a cycle was not completed for either inlet or outlet air streams. The MIBK concentration was lower in the outlet than in the inlet, indicating the plant uptaked MIBK. Water vapour concentration measured with PTR-MS was also almost constant for the inlet and outlet air streams. It was higher in the outlet due to the plant transpiration.

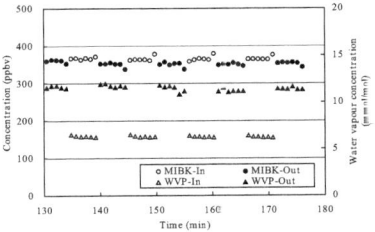

Fig. 3 Changes in MIBK and water vapour concentrations of the inlet and outlet air streams

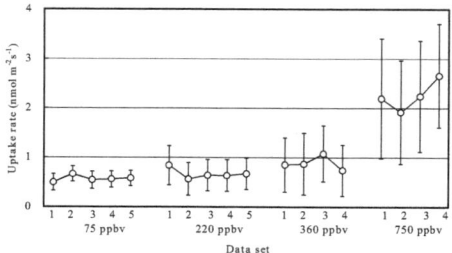

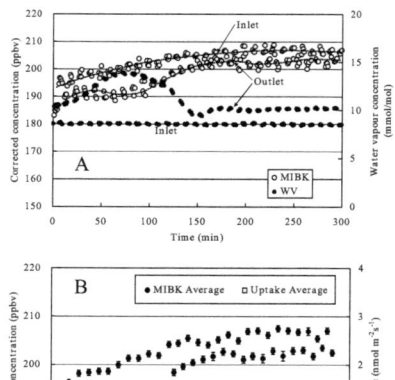

Fig. 4 MIBK uptake rate of golden pothos exposed to different concentration of MIBK. Bars indicate 95% confidence interval.

Fig. 4 shows MIBK uptake rate of golden pothos exposed to different concentration of MIBK. The uptake rate was calculated for individual outlet data sets and represented as an average±95% CL. Although the concentration difference between the inlet and outlet air streams was very small (4-9 %), the CL ranges were all positive, indicating MIBK uptake actually occurred on the plant leaf.

Measurement of dynamic change in plant's VOC uptake

Concentration changes in MIBK and water vapour for a relatively long term measurement

Fig. 5 Changes in MIBK and water vapour concentrations of the inlet and outlet air streams (A) and MIBK uptake rate of a golden pothos leaf (B).
Bars for MIBK concentration in (B) indicate SD and bars for the uptake rate in (B) indicate 95% confident interval.

is shown in Fig. 5. Because of a change in transpiration rate, water vapour concentration of the outlet air stream was unstable in the first half of the measurement. This means that stomata quite opened for the period of 70-110 min and then decreased its curvature, reaching a steady state in the latter half. A difference of MIBK concentration between the inlet and outlet air streams changed similarly while the outlet concentration was lower. This suggests that the MIBK uptake by the plant is regulated by the stomata opening. Uptakes of various ketone species by plants are shown in Table 1. All species of ketones treated here, except for acetone, were uptaked by plants and the total amounts uptaked by plants were always 1-2 orders higher than their amounts solubule into water contained in leaves. In this way, the dynamic change in plant's VOC uptake rate was monitored with the measurement system established here. The real-time monitoring using PTR-MS in this system might be a strong tool for measuring various VOC uptakes by plants.

Table 1 Ratio of intercellular to atmospheric concentrations (Ci/Ca) for various ketone species

Compound	Molecular weight	Henry constant (mol/kg*bar)	spathiphyllum		Pothos	
			Low concentration*	High concentration*	Low concentration*	High concentration*
Acetone	58.08	27	□	1.01±0.01	□	1.00±0.01
Methyl ethyl ketone	72.1	20	0.59±0.14	0.61±0.10	0.62±0.10	0.57±0.04
Diethyl ketone	86.13	20	0.61±0.04	0.48±0.09	0.61±0.05	0.58±0.06
Methyl propyl ketone	86.13	12	0.71±0.06	0.58±0.11	080±0.15	0.65±0.08
Methyl isobutyl ketone	100.16	2	0.75±0.11	0.78±0.09	0.71±0.01	0.80±0.07

*Low concentration (50□ 200ppb), High concentration (500□ 1000ppb)

PTR-MS measurements from surface sites in the UK and from the new UK BAe-146 research aircraft

David E. Oram[1], Anne Hulse[1], Stuart A. Penkett[1], and James R. Hopkins[2]

[1]*School of Environmental Sciences, University of East Anglia, Norwich, UK*
(d.e.oram@uea.ac.uk)

[2]*Department of Chemistry, University of York, York, UK*

A ground-based PTR-MS instrument has been operated in a semi-continuous mode at the Weybourne Atmospheric Observatory (WAO) since September 2002 to investigate the seasonal, diurnal and temporal behaviour of selected VOC and OVOC compounds in the atmosphere. Compounds measured include various oxygenated and aromatic hydrocarbons, isoprene, terpenes and DMS. WAO is situated on the North Norfolk coast in eastern England and experiences air masses from a variety of location with widely-varying levels of pollution (UK, continental Europe, North Sea). The instrument also took part in the UK TORCH (Tropospheric Organic Chemistry Experiment) experiment in the summer of 2003 where it was based at a more polluted site around 50 km downwind of London (Chelmsford, Essex). This experiment coincided with the high ozone event seen over much of Europe in August. A second PTR-MS has been adapted for use on the new UK Research Aircraft (BAe Systems 146, FAAM) and took part in the multi-national ICARTT (International Consortium for Atmospheric Research on Transport and Transformation) experiment in the summer of 2004. Flying out of Horta airport (Faial Island, Azores), the aircraft frequently intercepted plumes of pollution emanating from the North American continent, many of which were heavily impacted by the Alaskan forest fires. Preliminary data from both ground-based and airborne instruments will be presented together with some results from a comparison between PTR-MS and GC-FID techniques.

Atmospheric VOC Measurements at Thompson Farm and Appledore Island during the ICARTT 2004 Summer Campaign

Barkley C. Sive, Leif C. Nielsen, Robert Griffin, Ruth Varner, Yong Zhou, Pieter Beckman, Rachel Russo, Alex Pszenny, Bill Keene, Oliver Wingenter and Robert W. Talbot

University of New Hampshire, Institute for the Study of Earth, Oceans and Space, Climate Change Research Center, Durham, NH 03824 USA (bcs@ccrc.sr.unh.edu)

ABSTRACT

Two proton transfer reaction mass spectrometers (PTR-MS) were employed for atmospheric measurements of volatile organic compounds (VOCs) during the ICARTT summer campaign. One PTR-MS was located at the University of New Hampshire's Thompson Farm Observing Station (43.11 N, -70.95 W) and the second was located at Appleore Island (42.99 N, -69.34 W). Both instruments were run under the same conditions with a drift tube pressure of 2 mbar, an electric field of 600 V and a 20 second dwell time for measuring 24 VOCs. The resulting cycle time was approximately 10 minutes. Comparisons between the PTR-MS and other VOC measurements conducted on Appledore Island during the summer campaign are in good agreement overall. Other measurements on the island included mist chamber/ion chromatography measurements for acetic acid, canister samples analyzed by GC-ECD/FID/MS, and an in situ, cryo-less GC system for NMHCs and OVOCs. Trace gas distributions from both sites will be presented in addition to the local and regional impact on ozone production in New England from reactive VOC enhancements.

1. Introduction

One of the major research challenges in atmospheric chemistry is to understand fully the photochemical and dynamical processes that determine the rates of formation and loss of oxidants, such as ozone, throughout the atmosphere. Volatile organic compounds (VOCs) are involved in the photochemical production of tropospheric ozone and can directly influence hydroxyl radical (HO) concentrations. Oxidation processes affect the distribution and temporal trends of a large variety of trace gases emitted from natural and anthropogenic sources, including radiatively important trace gases such as methane. Ozone is also an important greenhouse gas [IPCC, 2001], but because of its short lifetime, its radiative impact is more difficult to assess than the well-mixed greenhouse gases. As a result, the feedback between chemistry and climate has potential implications for future global change. In order to understand these perturbations, high precision measurements of VOCs are essential.

The Atmospheric Investigation, Regional Modeling, Analysis and Prediction (AIRMAP) program is a University of New Hampshire (UNH) research program focused on atmospheric chemical and physical observations in rural to semi-remote areas of New Hampshire with the goal of understanding inter-relationships in regional air quality, meteorology, and climatic phenomena (http://airmap.unh.edu/). Continuous measurements of atmospheric constituents, including aerosols, ozone and its precursors are being conducted at several sites in New Hampshire to assess the air quality in this region. Instruments are located along a 200 km north-south transect in eastern New Hampshire beginning at near the Atlantic coast at Thompson Farm (Durham), moving up to 400 m elevation at Castle Springs (Moultonborough), and climbing to 2000 m at the summit of Mount Washington (North Conway). In addition to the three continuously operated AIRMAP sites, trace gases are also being measured during June-September on Appledore Island, Isles of Shoals. This site is located approximately 12 km off the New Hampshire coastline. Innovative weather forecasting technology is also being utilized to assess the synoptic meteorological patterns for this region. The relationships between local and far-field emissions, synoptic-to-local circulation, air quality, and New England climate variability of the past 100 years are also being explored through interpretation of spatial and temporal atmospheric observations and various modeling activities.

The International Consortium for Research on Transport and Transformation (ICARTT; http://www.al.noaa.gov/ICARTT/) is comprised of several groups from North America and Europe which planned the summer 2004 field experiments that were aimed at developing a better understanding of the factors that shape air quality in their respective countries and in the remote regions of the North Atlantic. The general objectives, goals and study areas of these programs overlapped significantly. Therefore, ICARTT was formed to take advantage of this synergy by planning and executing a series of coordinated experiments to study the emissions of aerosol and ozone precursors and their chemical transformations and removal during transport to and over the North Atlantic. The scale and scope of these capabilities have enabled unprecedented characterization of the key processes required to better understand and predict regional air quality, intercontinental transport, and radiation balance in the atmosphere by utilizing a large number of mobile platforms and ground stations. Presented here are measurements of NMHCs and OVOCs from two of the AIRMAP ground stations, Thompson Farm and Appledore Island, during the ICARTT summer campaign.

2. Measurements during the ICARTT Campaign

A PTR-MS from Ionicon Analytik has been used for fast response measurements of OVOCs, NMHCs and acetonitrile at the University of New Hampshire's Thompson Farm Observing Station since July, 2003. A second PTR-MS was deployed at Appledore Island for VOC measurements from July 1 through August 13, 2004, during the ICARTT summer campaign. Both instruments were run with a drift tube pressure of 2 mbar and an electric field of 600V while continuously stepping through a series of 30 masses. Of the 30 masses monitored, 6 masses were used for diagnostic purposes while the other 24 masses corresponded to the VOCs of interest. The dwell time for each of the 24 masses was 20 s during the ICARTT campaign, yielding a total measurement cycle of ~10 minutes. The systems were zeroed every 2.5 hours for 4 cycles by passing the flow through a catalytic converter (0.5% Pd on alumina at 425° C) in order to determine the system background signals.

Calibrations for the PTR-MS systems were conducted using three different high-pressure cylinders containing synthetic blends of selected NMHCs and OVOCs at the part per billion by volume (ppbv) level (Apel-Reimer Environmental, Inc.). Each of the cylinders used in the calibrations has an absolute accuracy of < ± 5% for all gases in each mix. Using volume dilution methods similar to those described by *Apel et al.* [1998a], the standards were diluted to atmospheric mixing ratios (ppbv to pptv levels) with whole air passed through the catalytic converter to scrub all VOCs and maintain the same humidity as the air being sampled. The calibrations were conducted regularly on both instruments to monitor their performance and to quantify the mixing ratios of the target gases. Additionally, mixing ratios for each gas were calculated by using the normalized counts per second which were obtained by subtracting out the non-zero background signal for each compound.

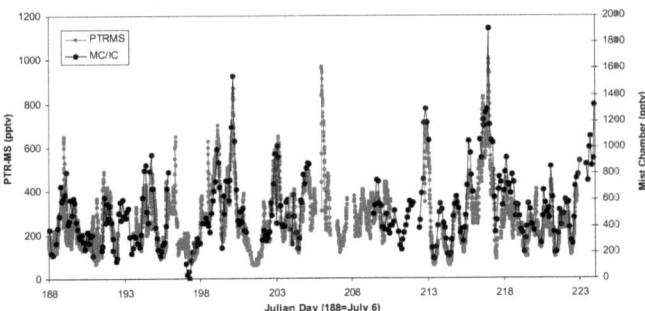

Figure 1. Time series plot of the preliminary data comparison between MC/IC and PTR-MS for acetic acid on Appledore Island from July 6 through August 9, 2004.

Comparisons between the PTR-MS and other VOC measurements conducted on Appledore Island during the ICARTT summer campaign are in good agreement overall. Other measurements that were deployed on the island included mist chamber/ion chromatography (MC/IC) for acetic acid, hourly canister samples analyzed by GC-ECD/FID/MS for NMHCs, halocarbons and alkyl nitrates, and an in situ, cryo-less GC system for NMHCs, OVOCs, halocarbons and alkyl nitrates. Figure 1 shows a time series plot of the preliminary acetic acid data from the MC/IC [A. Pszenny and W. Keene] and the Appledore Island PTR-MS. Overall, the MC/IC and PTR-MS measurements tracked each other remarkably well considering the fact that the PTR-MS was sampling from a 100' x 3/8" O.D. PFA Teflon line. The MC/IC and PTR-MS mixing ratios are plotted on separate y-axes to clearly illustrate that both measurements tracked each other, but that the calculated PTR-MS mixing ratios are about a factor of two lower than those of the MC/IC measurements. Given that we did not have an acetic acid standard during the summer campaign, the PTR-MS mixing ratios were calculated using the normalized counts per second. For the pressure and electric field used in the drift tube, this method can provide reasonably accurate mixing ratios without having to use calibration standards. However, for the Appledore Island PTR-MS, most calculated mixing ratios were lower than those determined from calibration standards. Because the calculated mixing ratios are largely dependant on an accurate transmission curve, at the end of the campaign, a new transmission curve for the Appledore Island PTR-MS was determined. Currently, we are investigating if the new transmission curve will bring the calculated mixing ratios in to better agreement with those of the calibration standards. Nonetheless, post-campaign calibrations of the 100' x 3/8" PFA Teflon line are being conducted in our laboratory to quantitatively determine the wall loss for acetic acid and the efficiency of passing ambient concentrations through the line under various atmospheric conditions. Ultimately, this will enable us to accurately quantify acetic acid mixing ratios for the ICARTT campaign.

Results from the PTR-MS are also compared with measurements conducted concurrently at Appledore Island using the cryo-less GC system (herein referred to as the MMR-GC) and from the whole air canister samples analyzed in our laboratory at UNH using a GC-FID/ECD/MS system. Overall, there is good agreement between all three measurement techniques for the subset of compounds that have been compared thus far. For the OVOCs, measurements from the MMR-GC system are compared to those of the PTR-MS. Figure 2 shows time series of plots of (a) acetone, (b) methanol, and (c) acetaldehyde mixing ratios from the MMR-GC and the PTR-MS. Both sets of measurements track each other extremely well and capture the small and large scale features. Additionally, absolute mixing ratios are in good agreement, with both yielding similar mean and median mixing ratios. For the PTR-MS acetone measurements, the influence of propanal was investigated to determine its contribution to the signal at mass 59. Both acetone and propanal were quantified on the MMR-GC system, and the propanal mixing ratios were found to be on the order of a few percent or less of the acetone mixing ratios. Therefore, we conclude that the dominate signal at mass 59 is from acetone and the contribution from propanal is negligible for this time period and location. Finally, it is worth noting that the largest discrepancies between these two systems occurred with measurements of acetaldehyde. Although mixing ratios are in agreement and both systems track each other reasonably well, there are discrepancies between the two systems during specific events where the PTR-MS mixing ratios are higher than those of the MMR-GC system. Further investigation of both systems is necessary in order to conclusively resolve this discrepancy.

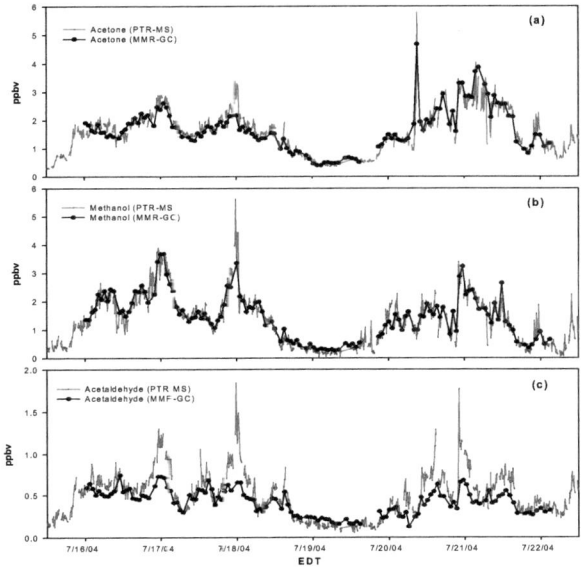

Figure 2. Time series plots of (a) acetone, (b) methanol, and (c) acetaldehyde for the MMR-GC and PTR-MS on Appledore Island from July 16-22, 2004.

3. Diurnal Cycles at Thompson Farm and Appledore Island

The diurnal cycles for propane and acetone at (a) Thompson Farm (TF) and (b) Appledore Island (AI) are illustrated in Figure 3 for August 2-4, 2004. The observations at TF are strongly coupled with unique boundary layer dynamics causing an extremely low-level nocturnal inversion allowing for pollutant buildup at night in this region. As seen in Figure 3a, a strong anti-correlation exists between propane and acetone at TF, while at AI, both gases tend to track each other fairly well (Figure 3b). On August 3 at TF, propane levels increased from a few hundred pptv during the day to approximately 4 ppbv at night. Furthermore, acetone levels dropped from approximately 4 ppbv during the night to ~1 ppbv by the early morning. The steady decrease in acetone at TF was most likely due to dry deposition since the reaction rates with NO_3 are relatively slow (i.e., on the order of 10^{-15} to 10^{-16} cm^3 molecule^{-1} s^{-1}) [*Akinson*, 1994]. Moreover, the mixing ratios of NO_3 are presumably kept low due to $NO_3 + NO_2 \rightarrow N_2O_5$, averaging a few hundred pptv in eastern New England during the summertime [*Brown et al.*, 2004]. In contrast to the diurnal trends at TF, both acetone and propane levels increased by the early morning on August 3 at AI (Figure 3b), illustrating how the shallow nocturnal inversion layer impacts the trace gas distributions over the continent. Using the observed decreases and increases for acetone, its depletion and replenishment rates have been estimated. At TF, day/night ratios for acetone averaged 2-3 while its deposition velocity was on the order of 1-2 cm s^{-1}, which is significantly higher than values used in many models. Such efficient removal may have important implications for the chemical impact of OVCCs, at least on a regional scale. Additionally, the large propane enhancements observed regularly at TF will be evaluated to determine the impact on local acetone production in the morning during periods of rapid replenishment. Finally, these gases will be used to demonstrate the importance of vertical mixing in driving the diurnal cycle of ground-level O_3 in New England.

31

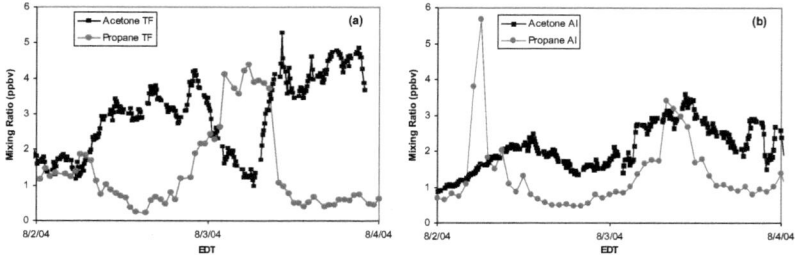

Figure 3. Time series plots of propane and acetone at (a) Thompson Farm and (b) Appledore Island from August 2-4, 2004.

4. Acknowledgements

We would like to thank Kevin Gervais and Don Blake from UCI for loaning us canisters for Appledore Island. Additionally, we wish to recognize the effort of the entire VOC group that helped make the summer campaign a great success. Financial support for this work was provided through the Office of Oceanic and Atmospheric Research at the National Oceanic and Atmospheric Administration and the National Science Foundation.

5. References

Apel, E. C., J. G. Calvert, J. P. Greenberg, D. Riemer, R. Zika, T. E. Kleindienst, W. A. Lonneman, K. Fung, E. Fujita, Generation and validation of oxygenated volatile organic carbon standards for the 1995 Southern Oxidants Study Nashville Intensive, *J. Geophys. Res.*, *103*, 22,281-22,944, 1998a.

Atkinson, R. , Gas-phase tropospheric chemistry of organic compounds, *J. Phys. Chem. Ref. Data,* Monograph No. 2, 1-216, 1994.

Brown, S. S., et al., Nighttime removal of NO_x in the summer marine boundary layer, *Geophys. Res. Lett.*, *31*, L07108, doi:10.1029/2004GL019412, 2004.

Climate Change 2001: The Scientific Basis, Third Assessment Report of the Intergovernmental Panel on Climate Change (IPCC).

On-Road Measurement of Vehicle VOC Emissions using PTR-MS

Berk Knighton, Todd Rogers, Eric Grimsrud

*Department of Chemistry and Biochemistry, Montana State University, Bozeman,
Montana 59717-3400 USA. bknighton@chemistry.montana.edu*

ABSTRACT

In the spring of 2003 (April 1-May 5), a multinational team of experts conducted an intensive
five-week field campaign in the Mexico City Metropolitan Area (MCMA). The overall goal
of this effort was to contribute to the understanding of the air quality problem in megacities.
As part of the campaign the Aerodyne Mobile Laboratory was equipped with state-of-the-art
analytical instruments and deployed for measuring a variety of vehicle emissions in real time
including CO_2, NO_2, NH_3, HCHO, VOC's and volatile (at 600 °C) aerosol. The on-road
measurement of vehicle VOC emissions were performed using a commercial version of the
IONICON PTR-MS modified to operate onboard the mobile lab platform. A significant
portion of the MCMA 2003 field campaign was dedicated to operating the mobile lab in the
"mobile chase mode" for the purpose of characterizing the emissions from specific classes of
vehicles operating under real world driving conditions. Diluted vehicle exhaust plumes are
intercepted by following or chasing the target vehicle. CO_2, a major exhaust product serves as
in internal tracer defines the duration and intensity of the intercepted plumes. The covariance
between the CO_2 and exhaust gas pollutant signals is used to define the presence of the target
analyte as a vehicle emission product. Selected chase events are presented to illustrate the
utility of the PTR-MS technique for characterizing vehicle VOC emission profiles in real
time.

1. Introduction

The Mexico City Metropolitan Area (MCMA) has been historically plagued by
degraded air quality. A significant source of air pollution in the MCMA arises from the
nearly 3 million motorized vehicles operating in the area. Emission inventory estimates
indicate that the transportation sector accounts for 80 % of NOx, 98% of CO and 40% of the
nonmethane VOC's (1). These emission inventory estimates are recognized to have
considerable uncertainty and that better characterization of the mobile source emissions is
needed. As part of the 2003 MCMA field campaign the Aerodyne Mobile Lab was deployed
to characterize emissions from in-use vehicles.

2. Description of the Aerodyne Mobile Lab

The Aerodyne Mobile Lab illustrated in Fig. 1 houses a combination of state of art
research grade and commercial instrumentation which have sufficient time response and
sensitivity needed to characterize dilute exhaust plumes. On-road vehicle emission
characterization was performed by driving in heavy traffic corridors or "chasing" curbside
commercial vehicles along their routes. In either mode of operation the partially diluted
exhaust plumes are sampled through a common sample inlet that protrudes through the
bulkhead of the truck directly above the driver. This flow (20 SLM) was isokinetically split
to provide flows to the particle and gas phase instruments. Approximately every 30 minutes
the sample line was flooded with pure nitrogen to provide instrument zeros and a time
synchronization event. A full description of the instrumentation deployed on the Aerodyne
Mobile Lab has been previously described (2) and only the PTR-MS will be discussed here.

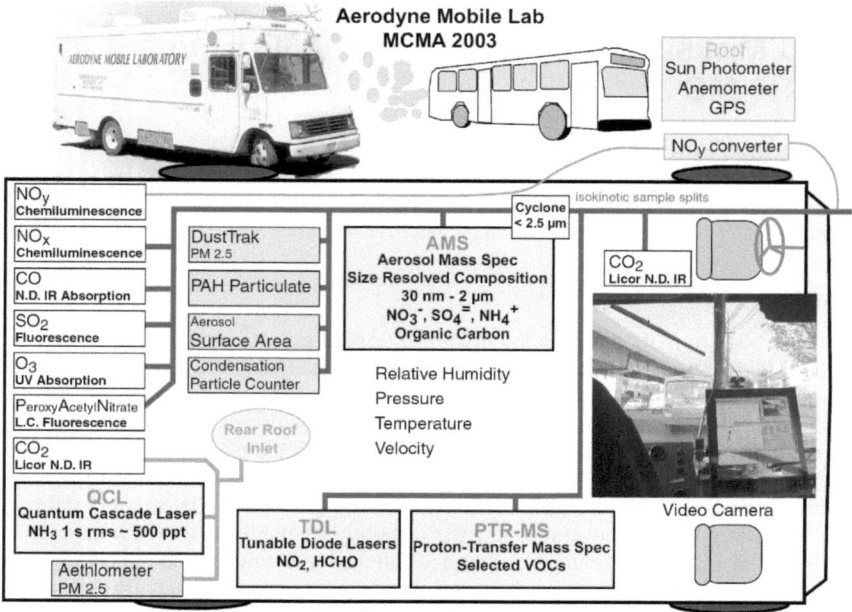

Figure 1. Schematic of the Aerodyne Mobile Laboratory.

3. Description of the PTR-MS

A commercial version of the PTR-MS (IONICON) was modified to handle the high impact and thermal loads experienced in MCMA field campaign. In addition to shock mounting the instrument rack and the drift tube mass spectrometer assembly, the orientation and manufacturer of the 210 L/s turbo pump were changed. A 250 L/s Varian Navigator 301 turbo pump was mounted vertically by the use of a 90° elbow. The magnetic bearing technology employed in the originally installed turbo pump was unable to handle the high impact shocks experienced within the mobile lab and had to be replaced with a fixed bearing style turbo pump. The small 33 L/s turbo pump handled the road shock without incident, but required a reduction in its rotational speed in order to handle the high internal mobile lab temperatures.

The original sample inlet flow controller was replaced by a pressure controller with a total flow rate of approximately 300 sccm. The inlet line was constructed using small diameter PFA teflon tubing and fittings except the stainless steel fine metering valve used to create the critical pressure drop. Linear velocity calculations indicate that the sample transit time through the inlet line is very short compared to the residence time within the drift tube. The drift tube pressure was nominally 1.9 mbar at 22°C and flow calculations predict a sample residence time of about 1.5 seconds within the drift tube.

For the on-road measurements 12 masses at 0.1 seconds per mass were monitored along with the drift tube pressure and temperature. The resulting duration of each measurement cycle was approximately 2 seconds. In addition to measuring the reagent ions H_3O^+ (m/z 21) and $H_3O^+(H_2O)$ (m/z 39) the following ions were monitored: m/z 33 – methanol, m/z 45 – acetaldehyde, m/z 57, m/z 59 – acetone, m/z 61 ethyl acetate fragment, m/z 79 benzene, m/z 89 ethyl acetate, m/z 93 toluene, m/z 107 – C2-benzenes and m/z 121 – C3-benzenes. Ambient air was sampled for 600 measurement cycles followed by 60 cycles of instrumental background determined by diverting the sample flow through a heated Pt catalyst that provided a VOC free gas stream.

4. Calibration and Concentration Determination

Trace gas concentrations were determined from either measured calibration factors or from theory. Calibration factors were determined from measured responses of a calibrated gas standard using equation 1 and are summarized in Table 1.

$$\left(\frac{I_{MH^+}}{I_{H_3O^+} + x \bullet I_{H_3O^+(H_2O)}} \right) = S_m \bullet \left(\frac{P_x^2}{T_x^2} \bullet X_m \right) \quad Eq\, 1$$

The pressure and temperature terms in equation 1 compensate for the dependence of these terms on the number density and reaction time making the sensitivity factor (S_m) independent of these variables. Table 1 also includes the sensitivity factors calculated as counts per second per ppbV normalized to 1 million cps reagent ion (ncps) at 2.4 mbar and 298 K. These values compare well to those reported in the literature by de Gouw et al. (3). Inter comparison of the PTR-MS derived concentrations with those determined from GC/FID measurements for the aromatic compounds at collocated sampling sites during the 2003 field campaign were in good agreement. The PTR-MS benzene measurement has been corrected for contributions from the fragmentation of ethyl and propylbenzene.

Table 1 PTR-MS calibration factors

Compound	S_m	x	Calc (ncps) 2.4 mbar	Lit.[3] (ncps) 2.4 mbar
Methanol	0.28	0.4	18.0	23.6
Acetaldehyde	0.73	0.6	45.0	26.6
Acetone	1.05	0.7	64.4	64.0
Benzene	0.51	-0.4	31.8	33.8
Toluene	0.72	-0.4	44.5	45.4
p-xylene	0.80	-0.4	49.8	18.7
1,2,4-trimethylbenzene	0.75	-0.4	46.4	30.2

5. Vehicle chase experiments

A sample chase event involving a colectivo microbus as it picks up and drops off curbside passengers is illustrated in Figure 2 where the CO_2, total aromatic (sum of benzene, toluene, C2-benzenes and C3-benzenes) and the ion detected at m/z 57 signals are plotted. Note the dramatic increase in the intensity of the m/z 57 signal and its excellent correlation with the CO_2 signal over the duration of this chase event. The temporal profiles of the m/z 57 and CO_2 signals clearly illustrates that the ambient air sampled by the mobile lab during this chase is reflecting the exhaust signature of the vehicle being followed. The total aromatic signal observed over the highlighted chase period is clearly different from than that of m/z 57 signal and the gasoline exhaust plumes bracketing this chase event. This vehicle was initially thought to be a gasoline vehicle but the lack of a significant aromatic signal indicates that this vehicle is probably using LPG fuel. For this vehicle the ion at m/z 57 is probably reflecting the presence of butene in the exhaust plume. In these experiments CO_2, a major exhaust product, serves as in internal tracer where the measured change in the m/z 57 and CO_2 signals within the plume yield a molar emission ratio. In this experiment the emission ratio is determined by the slope of the scatter plot of the m/z 57 versus CO_2 signal intensities. Molar emission ratios can be converted to emission factors (pollutant emitted/kg fuel) assuming nominal fuel combustion stoichiometry.

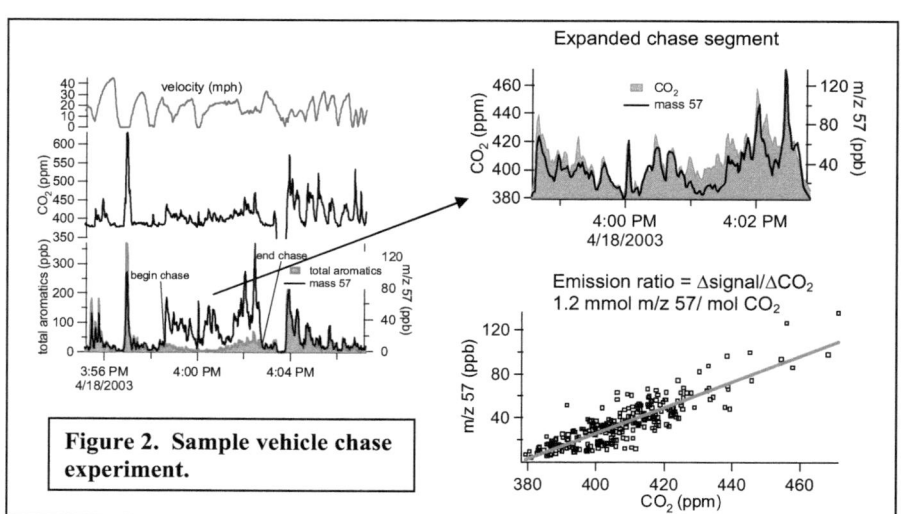

Figure 2. Sample vehicle chase experiment.

6. References

1. Air Quality in the Mexico City Megacity, An Integrated Assessment, L. T. Molina and M. J. Molina eds., 2002, Klewer, 253.
2. Kolb, C. E.; Herndon, S. C.; McManus, J. B.; Shorter, J. H.; Zahniser, M. S.; Nelson, D. D.; Jayne, J. T.; Canagaratna, M. R.; Worsnop, D. R., *Environ. Sci. Tech.* **2004**, in press.
3. de Gouw, J. A.; Goldan, P. D.; Warneke, C.; Kuster, W. C.; Roberts, J. M.; Marchewka, M.; Bertman, S. B.; Pszenny, A. A. P.; Keene, W. C., *J. Geophys. Res.* **2003**, 108, D21, doi:10.1029/2003JD003863.

Measurement of gas phase and aerosol composition in a smog chamber with a PTR-MS

J. Dommen[1], A. Gascho[1,2], M. Steinbacher[1,2]

[1]Paul Scherrer Institut, Laboratory of Atmospheric Chemistry, CH-5232 Villigen, Switzerland (josef.dommen@psi.ch)

[2]Swiss Federal Laboratories for Materials Testing and Research (EMPA), Air Pollution / Environmental Technology, CH-8600 Dübendorf, Switzerland

ABSTRACT

Many volatile organic compounds like terpenes and aromatics have been found to produce secondary organic aerosols (SOA). The nucleation process and the composition of these SOAs is still not well known. The inlet system of the proton-transfer-reaction mass spectrometer was modified to allow for the measurement of the gas and particulate phase of an aerosol. To measure the particulate phase the gaseous pollutants are stripped off and the particles are evaporated in a heated tube before entering the drift tube. SOA were produced during the photoxidation of 1,3,5-trimethylbenzene with NO_x in a smog chamber. This new setup allowed the observation of several mass peaks in the particulate phase.

1. Introduction

Ambient aerosol particles have a variety of important impacts, including adverse health effects, visibility reduction and their influence on climate. Carbonaceous aerosols are a major component of the total aerosol mass and can account for up to 50% of the fine particulate mass concentration ($PM_{2.5}$) on an annual average. However, their concentrations, size distribution and formation mechanisms are less understood than those of other compounds such as sulfate and nitrate.

The formation of secondary organic aerosols (SOA) has recently received much attention. Many natural volatile organic compounds (VOC) like terpenes, sesquiterpenes, isoprene as well as aromatic VOCs, which are mainly of anthropogenic origin, have been found to be precursors of SOA. Aromatics are emitted by fuel combustion and evaporation and will therefore influence SOA formation mainly in polluted urban areas. SOA is formed when the biogenic and anthropogenic precursor species react with atmospheric oxidants such as hydroxyl radicals, ozone, and nitrogen oxide species. Even in urban areas, up to 90% of the total organic aerosol mass can be attributed to SOA (Lim and Turpin, 2002).

The composition of SOA on the molecular level is difficult to analyze, and studies conducted so far were only able to resolve a small fraction of the entire organic particle mass (Seinfeld and Pandis, 1998). Many different oxygenated compounds like aldehydes, ketones and organic acids were observed in the particle phase. Over the last years numerous publications reported ambient gas phase measurements of alkenes, aromatics, alcohols, carbonyls and organic acids using Proton–transfer-reaction mass spectrometry (PTR-MS). Therefore, PTR-MS has the potential to measure oxygenated organic compounds not only in the gas but also in the particle phase. We developed a modified inlet system for the PTR-MS device which should allow us to measure the chemical composition of both phases. We present the instrument modifications as well as first measurements with this system at the smog chamber of the Paul Scherrer Institut.

2. Smog chamber and Instrumentation

The chamber is a 27-m^3 (3×3×3 m) flexible bag made of 125 μm (5 mil) DuPont™ *Teflon*® fluorocarbon film (FEP, type 500A, Foiltec GmbH, Germany). The bag is suspended in a temperature controlled wooden enclosure having dimension 4×5×4 m (L×W×H). The walls and ceiling of the enclosure are covered with reflective aluminum foil to maximize the light intensity and increase light scatter or diffusion. The aluminum foil has greater than 80% reflection for light with wavelength greater than 300 nm. The chamber temperature is controlled by two cooling units allowing for temperature stabilization of ±1°C within the range of 15 to 30°C. Two manifolds (inlet and outlet) made of stainless steel and *Teflon*® allow for easy installation of additional inputs and sampling lines. Four xenon arc lamps (4 kW rated power,1.55×10^5 lumens each, XBO® 4000 W/HS, OSRAM) are used to simulate the solar light spectrum and to mimic natural photochemistry.

There are many instruments attached to this smog chamber to measure the physical parameters of the particles and their chemical composition as well as the gas phase precursors and oxidized product compounds. Results are presented from the Environics S300 Ozone Analyzer (UV-photometer), Thermo Environmental Instruments 42C Trace Level NO$_x$ analyzer and the Aero Laser 5002 carbon monoxide monitor.

For the gas phase measurements shown here we operated the PTR-MS instrument (IONICON Analytik GmbH, Innsbruck, Austria) without modifications and as described in detail by Steinbacher et al. [2004]. To measure molecular constituents of particles in an aerosol we modified the inlet system of the instrument. First, the gas phase is stripped off by a charcoal denuder. Then the aerosols are thermally evaporated in a coiled stainless steel tube heated to 200 °C. The existing needle valve and the relatively long capillary were replaced with orifices and very short tubes to minimize losses of organic compounds.

3. Measurements

To test the performance of the PTR-MS instrument and the new inlet system experiments in an environmental reaction chamber were used to simulate the photochemical degradation of a hydrocarbon and to generate organic particles. The chamber is humidified to 50% nominal relative humidity. Next, NO, NO$_2$, and propene are flushed in sequentially. A small amount of a liquid hydrocarbon is injected into a heated glass bulb which is then flushed into the chamber with pure air. Finally, the contents are left to mix for approximately 45 minutes before turning on the lamps. The experiments presented here were conducted with initial input concentrations of 1160 ppb 1,3,5-trimethylbenzene (1,3,5-TMB), 240 ppb NO, 240 ppb NO$_2$, and 300 ppb propene.

The photochemical degradation of 1,3,5-TMB produces a large number of oxidation products, which can be observed with PTR-MS. Figure 1 shows a mass spectral scan after 2.5 hours of irradiation in the reaction chamber. Products with masses up to m/z of 185 can be observed. The main problem is the identification and quantification of these mass signals. Some mass peaks were assigned based on previously reported observations in literature as well as other measurement techniques like ion chromatography and GC-MS (after derivatization) operated at our chamber in parallel to PTR-MS. Even for known mass peaks quantification is difficult due to lack of gaseous calibration standards of those compounds. In those cases the reaction rate constants are calculated by the theory of Su and Chesnavich [1982]. In all other cases a standard rate constant of $2\cdot10^{-9}$ cm^{3}molecule^{-1}s^{-1} has been chosen. Figure 2 shows the evolution of some precursors and oxidation products over 8 hours of irradiation. The precursors NO and 1,3,5-TMB are consumed and decrease steadily. Ozone reaches a maximum after about 3 hours. One of the main oxidation products is methylglyoxal (m/z 73), which reaches the maximum concentration after 3.5 hours and then decreases due to photolysis and reaction with OH radicals. CO mixing ratios are steadily increasing since it is produced from secondary and later generation oxidation products.

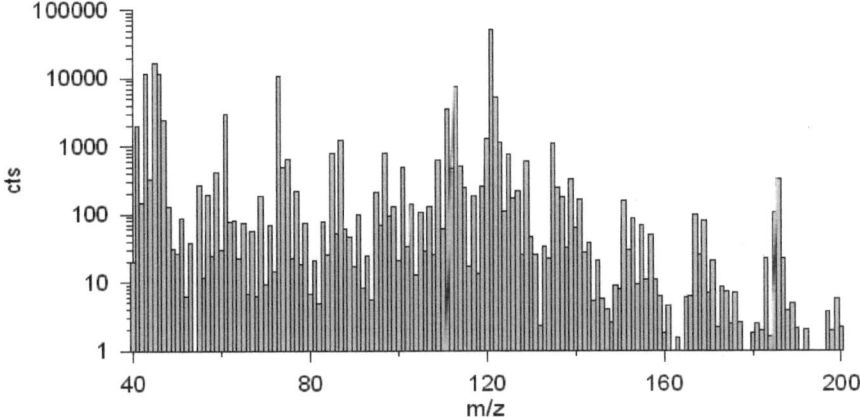

Figure 1: Mass scan of gas phase in reaction chamber after 2.5 hours (average of data between 2 and 3 hours) of irradiation of a 1,3,5-TMB, NO$_x$, propene mixture (see text).

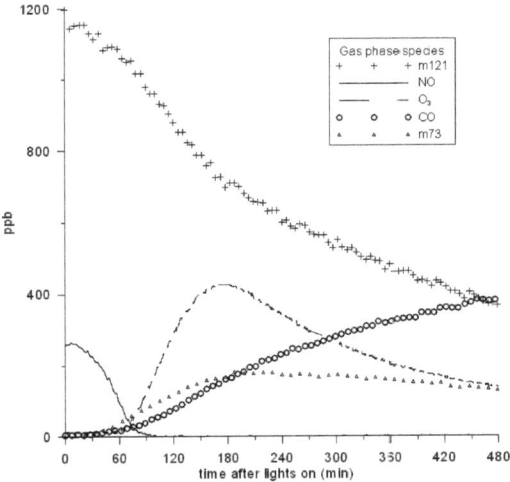

Figure 2: Mixing ratio versus time for O$_3$, NO, CO and the m/z 73 and 121, corresponding to methylglyoxal and 1,3,5-trimethylbenzene respectively, measured with the PTR-MS. Starting conditions: 1160 ppbv 1,3,5-trimethylbenzene. 1,3,5-trimethylbenzene was calibrated with a gas standard while the mass signal of methylglyoxal was converted to mixing ratios using the reaction rate constant $k_{73}= 2.05 \cdot 10^{-9}$ cm^{3}molecule^{-1}s^{-1}.

The formation of SOA starts slowly after about one hour and a rapid increase occurs after 2 hours of irradiation. The temporal development of the SOA mass is presented in Figure 3 together with some mass peaks observed with PTR-MS in the particulate and gas phase. The mass traces of the gas and particle phase are not from the same experiment, since the two phases can only be measured sequentially. Showing the full time development of each phase gives a clearer picture of the ongoing processes. As can be seen in Figure 3 the mixing ratios of the gas phase compounds m/z 87, 111 and 113 start to increase after lights on. However, the same compounds can not be observed in the particle phase inlet until SOA is formed. Then they increase in parallel to the SOA mass concentration. It should also be noted that m/z 113 already starts to decrease when SOA is strongly formed while in the particle phase m/z 113 increases with SOA mass.

These experiments demonstrate that it is possible to measure the gas as well as the particle phase of an aerosol with PTR-MS. To have an instrument able to measure both phases in an on-line mode and at such a high time resolution is unique.

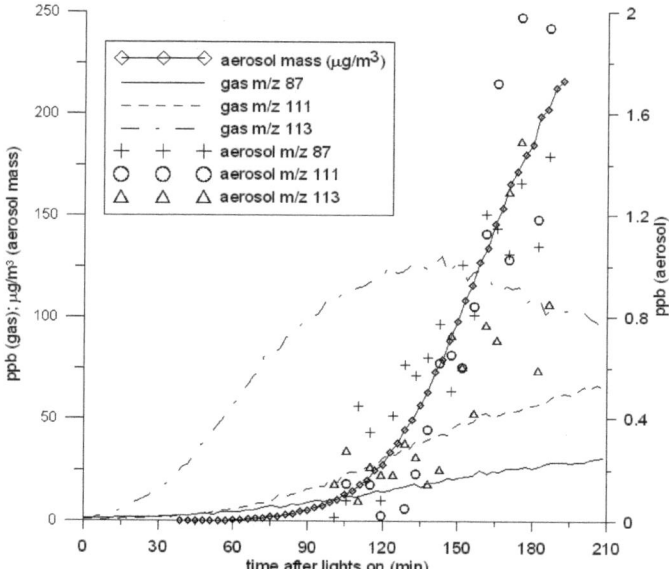

Figure 3: Temporal development of the aerosol mass in the reaction chamber as well as the mass peaks m/z 87, 111 and 113 measured in the particle phase with PTR-MS. Traces of the gas phase mass peaks of m/z 87, 111 and 113 are also shown from another similar experiment. Left y-axis: gas phase mixing ratios in ppb and aerosol mass in μg/m³, right axis: mixing ratio of mass peaks from particulate phase in ppb.

4. References

Lim H.-J., B.J. Turpin, Origins of primary and secondary organic aerosol in Atlanta: Results of time-resolved measurements during the Atlanta supersite experiment, *Environ. Sci. Technol.*, **36**, 4489-4496 (2002).

Seinfeld J.H. S.N. Pandis, *Atmospheric Chemistry and Physics of Air Pollution*, Wiley, New York (1998).

Steinbacher M., J. Dommen, C. Ammann, C. Spirig, A. Neftel, A.S.H. Prevot, Performance Characteristics of a Proton-Transfer-Reaction Mass Spectrometer (PTR-MS) derived from laboratory and field measurements, *Int. J. Mass Spectr.*, in press (2004).

Su T., W. J. Chesnavich, *J. Chem. Phys.* **76**, 5183 (1982).

Oxidation of Monoterpenes in the Presence of Aqueous Sulphate Particles

Ralf Tillmann, Thomas F. Mentel, Astrid Kiendler-Scharr

Institute for Chemistry and Dynamics of the Geosphere, ICG-II: Troposphere, Research Centre Juelich GmbH, Germany (r.tillmann@fz-juelich.de)

ABSTRACT

We investigated the partitioning of the ozonolysis products of monoterpenes between gas and particulate phase. The experiments were performed in the large Aerosol Chamber in Juelich. Sabinene and α-Pinene were chosen as reactands; aqueous ammonium sulphate served as seed aerosol. The gas-phase products were measured by PTRMS. The condensation of oxidation products onto the sulphate particles was monitored by AMS (Aerosol Mass Spectrometry). From the PTR-mass spectra fragmentation patterns and product masses were identified by linear regression analysis and by correlation analysis, respectively. We derived the fraction of carbon per consumed monoterpene which remained in the gas phase by attributing carbon numbers to the product masses. Similar, the fraction of carbon which appeared in the particulate phase was derived from the AMS-data. For the α-Pinene a total carbon recovery of ~0.6 (0.35 + 0.25) could be achieved for initial mixing ratios of 50 ppb O_3 and 40 ppb of α-Pinene. The heterogeneous N_2O_5 hydrolysis served as a tool to determine the hygroscopic properties of the particles after condensation of the monoterpene oxidation products..

1. Introduction

Biogenic volatile organic compounds (BVOC) like the monoterpenes ($C_{10}H_{16}$) are emitted to the atmosphere in large quantities by plants. During their oxidation, parts of the products may condense on atmospheric particles. If those oxidation products act as surfactants they can change the properties of the aerosols. For example, evaporation of water might be reduced by an organic film on the aerosol surface or the transport of hydrophilic molecules from the gas phase to the aerosol phase can be impeded. All processes acting on the chemical composition of the aerosol will be of immediate importance for light scattering properties and the ability to act as cloud condensation nuclei.

2. Experimental

Oxidation experiments of monoterpenes (Sabinene and α-Pinene) were conducted in the large Aerosol Chamber of the Research Centre Juelich. The chamber consists of a Teflon FEP bag of 260 m³ volume with a volume/surface ration of ~ 1. Cleaning is provided by flushing the chamber with ultra pure air. The aerosol is generated by spraying dilute salt solutions inside a small external chamber and flushing the fine fraction of aerosol into the aerosol chamber. The chamber is equipped with an FTIR-Spectrometer, Ozone-Monitor, Aerosol Mass Spectrometer (Aerodyne) and a high sensitivity PTRMS (IONICON). Aerosol distributions are monitored by means of an TSI-SMPS (Scanning Mobility Particle Sizer) and an TSI-APS (Aerodynamic Particle Sizer). For further characterization of the aerosol composition a SJAC (Steam Jet Aerosol Collector) combined with an Ion Chromatography system is available. The monoterpenes were added into the chamber by injecting the pure liquids with a μL-syringe after the aerosol was generated. Using 20 – 60 μL of pure monoterpenes resulted in mixing ratios of 13 – 40 ppb. Oxidation was initiated by the injection of ozone. The O_3 concentration was increased stepwise to suppress new particle formation. (The ozonolysis of monoterpenes produces OH-radicals (e.g. *Atkinson and Arey, 2003*), thus the oxidation is driven by O_3 and OH.) At the end of the oxidation process O_3 is topped up to a mixing ratio of ~ 1.7 ppm for the subsequent N_2O_5 hydrolysis. The O_3 is titrated by injecting NO_2 into the chamber and N_2O_5 is produced *in-situ*. The uptake of N_2O_5 is monitored by FTIR-gas phase measurements.

3. PTR-mass spectra analysis

For monitoring the gas-phase mixing ratios of the monoterpenes and their oxidation products the PTRMS was operated alternately in the single ion mode and the bargraph scan mode. Single ion detection was performed for masses lower than 50 amu with varying integration times. Above 50 amu up to 199 amu a sample time of 2 sec. per mass was chosen. Performing a correlation analysis of the educt mass with all other masses results in positive correlation coefficients for all fragments of the educt and its isotope signals. All products feature a negative correlation coefficient as long as the oxidation proceeds. We assume that all masses with a correlation coefficient lower than -0.9 are oxidation products (see Fig. 1). Masses with higher correlation coefficients but a pronounced time respond to the oxidation process were included in the data analysis (dotted cicles in Fig. 1). For the α-Pinene experiment this way 16 product ions were identified. If the ^{13}C signal (m+1) exceeds the limit of detection the C-number of the respective molecule can be determined by plotting the ^{13}C vs. ^{12}C for each time (see Fig. 2). The slope of the linear regression provides a value of the $^{13}C/^{12}C$ ratio. Considering a natural abundance of ^{13}C of 1.1% the C-number can be calculated.

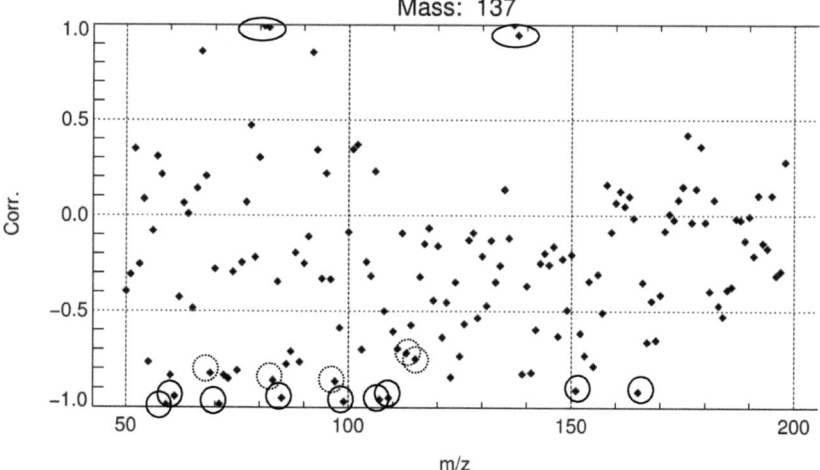

Figure 1: Correlation coefficient of the reactant mass signal 137 with all other mass signals during an α-Pinene oxidation experiment. The fragments of α-Pinene (m/z 81;137) and their isotope signals (m/z 82,138) are strongly positively correlated. The product mass signals are negatively correlated (circles).

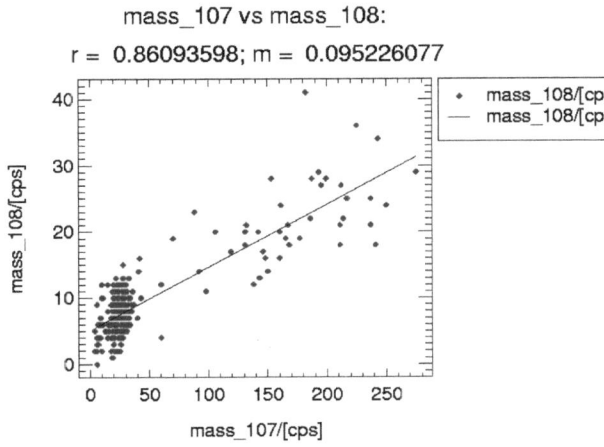

mass_107 vs mass_108:

r = 0.86093598; m = 0.095226077

Figure 2: Plot of mass 107 against its isotope mass signal. r denotes the correlation coefficient, m the slope of the linear regression fit. The slope has a value close to 0.099 as it would be expected for a C-9 molecule.

A calibration of the PTRMS for the monoterpenes and the identified low molecular weight compounds was performed using a diffusion source operated in the low ppb mixing ratio range. For mass signals which could not be identified and those molecules which could not be calibrated by means of a diffusion source, we assumed the sensitivity given for the monoterpene to calculate its mixing ratio.

Mixing ratios were then calculated as in terms of carbon-equivalents. The carbon mixing ratios of the identified products were added up and set in relation to the carbon mixing ratio of the consumed monoterpene. This is the carbon ratio specifying the recovery of carbon in the gas phase (see Fig. 3).

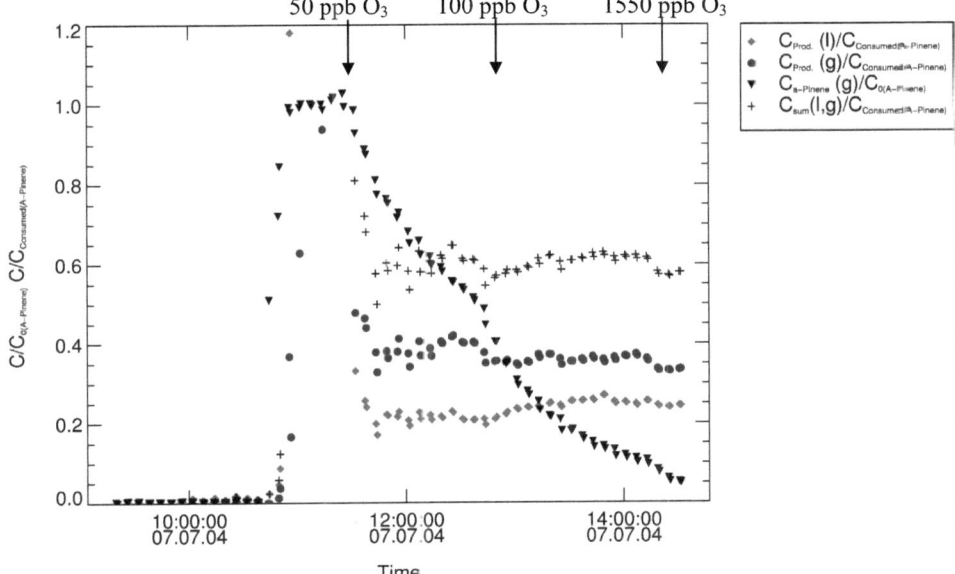

Figure 3: α-Pinene oxidation: α-Pinene is given normalized to its initial mixing ratio. The respective values for the oxidation products are added up and normalised to the C-mixing

ratio of consumed α-Pinene. Gas phase products and the monoterpene are measured by PTRMS, the particulate phase is measured by AMS.

4. Results and Discussion

The recovery of the gas-phase oxidation products of α-Pinene has been determined to >35% by PTRMS measurement. It tends to drop slightly to longer times. Considering fragmentation of high mass molecules inside the drift tube this recovery rate is expected to be a lower estimate. Furthermore the reaction product Formaldehyde was not quantified by PTRMS and is not included in the mass balance. Considering the yield of HCHO from the α-Pinene oxidation as stated by *Atkinson and Arey, 2003* the contribution of Formaldehyde to the carbon balance is at maximum 2.2 %, thus negligible. Wall losses, inlet line losses are a source of error for the quantification of the sticky molecules like the organic acids.
Aerosol mass spectrometry shows a slow increase of the organic component of the aerosol. By estimating a mean molecular weight of the aerosol phase products from literature data (*Yu et al., 1999*) we obtained the product-carbon/$C_{consumed}$ ratio for the particulate phase. The recovery of the condensed-phase oxidation products of α-Pinene has been determined to <25%. This value increases slowly with time.
In total the recovery rate adds up to ~60% and compares with the results of *Yu et al.*
The chemical characteristic of the aerosol was determined by N_2O_5-hydrolysis on the aerosol. It is evident that the increase of the organic fraction reduces the uptake coefficient of the N_2O_5 into the particulate phase (see tab. 1). Therefore the water of the inorganic aerosol is shielded significantly by the monoterpene oxidation products (see also *Folkers et al., 2003*).

Experiment	Uptake coefficient
60 µL α-Pinene NH_4HSO_4-Aerosol	$\geq 1.28 \ 10^{-3}$
20 µl Sabinen NH_4HSO_4-Aerosol	$1.16 \ 10^{-2}$
NH_4HSO_4-Aerosol	$2.0 \ 10^{-2}$

Table 1: Uptake coefficients of N_2O_5 on NH_4HSO_4-Aerosols

5. Conclusions

The oxidation of monoterpenes was observed in a large Aerosol Chamber. The degradation of the educts and the appearance of the oxidation products in the gas phase was monitored by PTRMS. Simultaneously, the formation of the organic aerosol component was observed by AMS. PTR mass spectra analysis by regression and correlation together with analogous AMS data delivers useful information for the calculation of the carbon conversion during monoterpene oxidations. We recovered a total of ~60% C, >35% in the gas phase and <25% in the particulate phase. The organic fractions in the particles retarded significantly the heterogeneous N_2O_5 hydrolysis. In the atmosphere this would slow down the removal of nitrogen oxides during the night.

6. Literature

Atkinson, R. Arey, J., Gas-phase tropospheric chemistry of biogenic volatile organic compounds: a review, *Atmos. Environ. 37 (S2)*: S197-S219, 2003
Folkers, M., Th. F. Mentel, and A. Wahner, Influence of an organic coating on the reactivity of aqueous aerosols probed by the heterogeneous hydrolysis of N_2O_5, *Geophys. Res. Let. 30 (12):* Art. No. 1644, 2003
Yu, J., Cocker III, D. R., Griffin, R. J., Flagan, R. C., Seinfeld, J. H., Gas-phase ozone oxidation of monoterpenes: Gaseous and particulate products, *J Atmos. Chem. 34*: 207-258, 1999

Calibration of PTR-MS for OVOCs

Shungo KATO[1], Yuko MIYAKAWA[1], Yoshizumi KAJII[1]

[1]Tokyo Metropolitan University, Minamiosawa1-1, Hachioji-shi, Tokyo 192-0397, JAPAN (shungo@atmchem.apchem.metro-u.ac.jp)

ABSTRACT

PTR-MS was calibrated for OVOCs (oxygenated volatile organic carbons) measurements. Their concentrations were determined by two method. Diffusion tubes and a pre-mixed cylinder. Response of OVOCs signals showed linearity for large concentration range. Their absolute values were stable more than 2 month when the PTR-MS keep running. It seems to change after the PTR-MS was totally stopped. Humidity dependence of formaldehyde was examined. Formaldehyde showed linear response as concentration increase at constant humidity condition, and decreased as humidity increase. This trend was explained by simple model calculation and the rate constant of backward reaction was estimated $3*10^{-11}$ cm^3 $molecules^{-1}$ s^{-1}.

1. Introduction

PTR-MS is quite convenient instrument to measure atmospheric species. It does not required carrier gas and special setting, and reports concentrations with high time resolution. PTR-MS also have an advantage that it can measure OVOCs, which contain oxygen such as alcohol, aldehyde, and ketone. Since some OVOCs have polarity and are sticky, it will need special care to measure them by GC methods. Now PTR-MS seems to be a very easy, convenient, and useful instrument, but we need to be careful with the results it produced. PTR-MS is only monitoring the targeted mass number, interference by same mass number species and fragment peaks. There are also other factors to change the "calculated concentration" by the principle equation of the proton transfer reaction. Therefore it is necessary to calibrate the targeted species. We usually take some air samples for GC measurements during PTR-MS measurements. The continuous PTR-MS results are correct by the results of some GC measurements. But in the case of OVOCs, the GC results is not enough trustful to calibrate the PTR-MS results. For OVOCs, we used standard gas before and after measurements for calibration. Here we will explain the method of calibration and problems. Formaldehyde has only slightly higher proton affinity than H_2O and backward reaction is taken place in the drift tube. Humidity dependence of formaldehyde was examined. In this article, some species which are not in the definition of OVOC are also called OVOC for convenience.

2. Experiments

Two calibration method were applied for OVOC measurements. Diffusion tubes and pre mixed cylinder. A diffusion / permeation tubes were used to generate known concentration of OVOCs. From the weight decrease of the OVOCs in the diffusion tube, the concentration can be calculated. The generated standard gas was diluted by nitrogen and measured by PTR-MS. Standard gases of methanol, ethanol, acetone, acetonitrile were prepared by diffusion tubes and of acetaldehyde was prepared by permeation tube.

6 species mixed cylinder (methanol 10ppm, ethanol 10ppm, acetaldehyde 5ppm, acetone 5ppm, isoprene 1ppm, acetonitrile 5ppm) were prepared by weight filling method (Taiyo-Toyo Sanso, Japan). Using two mass flow controllers (MFC), this standard gas was diluted by nitrogen and introduced to PTR-MS. To avoid the loss in the lines, teflon tubes were used as

short as possible and were heated if the room temperature was cold. It take long time to stabilize, especially for methanol, and we need to wait about half day.

Standard gas of Formaldehyde in cylinder (1 ppm) was prepared by Takachiho, Japan. This standard gas and were mixed by two MFCs. For humidity dependence experiment, only nitrogen flow was through a water bubbler and change the humidity of the flow.

3. Results and Discussion

The permiation tube makes stable concentration of OVOCs. Because of the limitation of the MFCs, only considerably high concentration of OVOCs can be injected to PTR-MS. But they show good linearity for methanol (m/z=33), acetaldehyde(m/z=45), acetone(m/z=59) ,and acetonitrile(m/z=42). But in the case of ethanol (m/z=47), the signal was much lower than expected and did not show linearity against the injected concentration. It seems to be impossible to measure ethanol by PTR-MS. We believe the results of using diffusion tubes are more reliable than the pre-mixed cylinder because the stability during storage. But in practically, calibration using by cylinder is much easier. The standard gas cylinder was calibrated by the diffusion results. When comparing the results, the concentrations of the cylinder were about 30% lower than the results of diffusion tubes. This comparison would be required periodically because the stability in the cylinder. The sensitivity of the PTR-MS for OVOCs were stable more than 2 month if it keep running. But we calibrate for OVOCs before and after the measurements campaign. The sensitivity of the PTR-MS seems to be change when the system shutdown (vacuum system was stopped). That means the calibration is required when the system stopped.

Formaldehyde (m/z=31) showed good linearity when the injected concentration changed using two MFCs. This trend was same at constant humidity. In figure 2 the relative signal change at various humidity was plotted. It can be seen that when the humidity was increased, the signal was decreased. The rate constant of the backward reaction ($HCHOH^+ + H_2O \rightarrow HCHO + H_3O^+$) was assumed and the expected signal decrease was calculated. When the rate constant of the backward reaction was assuming $3*10^{-11}$ cm^3 molecules^{-1} s^{-1}, the experimental results is fitted well. This rate constant is similar to the reported value by Hansel et al. ($2*10^{-11}$ cm^3 molecules^{-1} s^{-1}, IJMS167/168, 697-, 1997). In typical ambient air condition ($H_2O = 1.5*10^{15}$ molecules cm^{-3}), the formaldehyde signal will decrease about 60%. When using suitable correction by humidity, formaldehyde concentration could be obtained by PTR-MS measurements.

On-line Determination of Deuterium Abundance in Water Vapour Using a Proton Transfer Reaction – Mass Spectrometer

Sean Hayward*, Michael Wilkinson and C. Nicholas Hewitt

Institute of Environmental and Natural Sciences, Lancaster University, Lancaster, LA1 4YQ, U.K.

*Current Address: Geocentrum II, Sölvegatan 12, 223 63, Lund University, Lund, Sweden. Email: sean.hayward@nateko.lu.se

ABSTRACT

This paper presents the application of PTR-MS to the measurement of deuterium in the headspace vapour of D_2O spiked water. For this application, the PTR-MS is operated with a much lower electric field strength than normal promoting reagent ion (RI) hydration by water vapour. The hydrated RI spectrum is dominated by the $H_3O^+(H_2O)_3$ cluster (m/z 73) and its associated ^{17}O, ^{18}O and deuterated isotopic variants (IVs). The relative abundances of these IVs are 'isotopically amplified' when compared with the dominant IVs of the smaller unhydrated RI. This amplification is crucial to the precise analysis of deuterium in water vapour by PTR-MS.

By adopting the known fractional abundances of ^{16}O, ^{17}O, ^{18}O, ^1H and D for each IV of the m/z 73 hydrated cluster, we show how measurement of the ratio of m/z 74 to m/z 75 enables us to accurately determine the deuterium content of water vapour. We conclude by determining total body water content by breath analysis following ingestion of D_2O.

1. Introduction

It was recently shown that the fractional deuterium (D) abundance of water vapour, and the head-space vapour of water spiked with small quantities of D_2O ($H_2O + D_2O \rightarrow 2HDO$) could be determined on-line using selected ion flow tube and flowing afterglow - mass spectrometry (SIFT/FA-MS)[1]. FA-MS was subsequently applied to the determination of deuterium in the water vapour of human breath, enabling the accurate determination of total body water (TBW) in patients following ingestion of a small quantities of pure D_2O[2,3]. These techniques rely upon isotope exchange reactions of the tri-hydrated clusters of H_3O^+ reagent ions with water vapour molecules (H_2O, HDO, $H_2^{17}O$ and $H_2^{18}O$) at thermal energies within the ion flow tube. The exchange reaction of the tri-hydrate cluster of the H_3O^+ reagent ion with HDO is given in Equation 1;

$$H_3O^+(H_2O)_3 + HDO \leftrightarrow H_8DO_4^+ + H_2O \qquad (1)$$

Tri-hydrated cluster ions and associated isotopic variants (IVs) are formed from H_3O^+ reagent ions and water molecules and have m/z 73; $H_3O^+(H_2O)_3$, m/z 74; $H_8DO_4^+$ and $H_9{}^{17}OO_3$ and m/z 75; $H_9{}^{18}OO_3^+$. Ions are counted at m/z 74 and 75, the former as a direct measure of singular D incorporation into the cluster, the latter as a surrogate measure for the dominant m/z 73 cluster. By adopting the known fractional abundances of ^{16}O, ^{17}O, ^{18}O, ^1H and D for each species, and their corresponding liquid/vapour phase partition coefficients, the ICR ratio of m/z 74 to m/z 75, Q, yields information on the deuterium content of the vapour being analysed[1]. From this, the deuterium content of the liquid phase is easily determined.

PTR-MS has two significant operational differences to the techniques of SIFT/FA-MS which must first be addressed before deuterium measurements can be considered using the theory outlined above. First, PTR-MS is typically operated with an electric field strength of 66 V cm^{-1} and at a pressure of 200 Pa, giving an E/N ratio of 120-130 Td. Within this range, the desired tri-hydrated cluster ions are largely suppressed. Second, in PTR-MS, there is a small 'bleed' of water vapour from the ion source to the drift tube which dilutes the water vapour in the sample air. Here, we describe how the two aforementioned challenges are overcome enabling the first on-line determination of deuterium in water vapour using a PTR-MS. We conclude by repeating Spanel and Smith's deuterated water tracer experiment for the determination of TBW.

2. Method

Experiments were conducted using a standard PTR-MS. The drift tube was maintained at 280-300 Pa and 80°C, whilst an electric field of 18 V cm^{-1} was applied along the length of drift tube giving an E/N ratio of 27-29 Td. Under these conditions, the ICR of the m/z 73 reagent ion water cluster was stable, dominating the reagent ion water cluster spectrum, whilst the water vapour bleed was partially suppressed. The typical ICR of m/z 73, determined indirectly by measurement of the m/z 75 ^{18}O IV was typically 3.5×10^6 cps for water vapour in human breath (i.e. 37°C).

Sample air was delivered to the instrument at ~100 mL min^{-1} via a heated (80°C) 0.32 mm i.d. PFA transfer line, and a sub-sample introduced to the instrument via a heated 0.16 mm i.d. 'silcosteel' capillary line. The lag time between water vapour source and PTR-MS detector was < 1 s. Ion source water vapour was produced from deionised tap-water. The flow of water vapour into the ion source was regulated by mass-flow controller and in most cases was 6 sccm.

Initial experiments focussed on the head-space vapours of standard reference mixtures of D_2O and tap-water bubbled with dry air. To develop a correction term for the impact of ion source water vapour bleed, Q was determined for a 2004 ppm standard reference mixture for ion source water vapour flows in the range 0 - 6 sccm. To confirm the accuracy of the technique, including the application of the correction term, the response of the instrument in the range 154 ppm (tap-water) – 2004 ppm deuterium was assessed.

Finally, we repeated the deuterated water tracer experiment of Spanel and Smith (2000) for the determination of total body water. A male volunteer of known mass consumed ~30 mL of D_2O diluted in ~200 mL tap-water. At the same moment, another male volunteer consumed a similar quantity of tap-water only. For 20 minutes prior to, and 140 minutes post-ingestion, Q was periodically monitored for the breath water vapour of both volunteers. Exhaled breath samples were introduced into the PTR-MS inlet *via* a disposable mouth-piece located in the end of a heated (80°C) acrylic sample tube (15 mm i.d. × 1 m length) into which the PFA transfer line from the PTR-MS was loosely fed.

3. Results and Discussion

Figure 1a) presents Q, the *m/z* 74 to *m/z* 75 ICR ratio for the headspace vapour of 2004 ppm deuterated water, as well as the ICRs of *m/z* 74 and 75, as water flow into the ion source is varied. From the linear slope, the diluting effect of ion-source water bleed on the ratio Q can clearly be seen. Analysis of this slope enabled an empirical water-bleed correction expression to be developed, enabling measurements made at 6 sccm ion source water flow to be corrected to zero-water flow. Figure 1 b) shows the subsequent determination of seven standard reference mixtures, between 154 ppm deuterium (i.e. tap-water) and 2004 ppm. Clear linearity is observed with all data points lying within ± 2.5% of unity. Relative standard deviation for each data point is less than 1%.

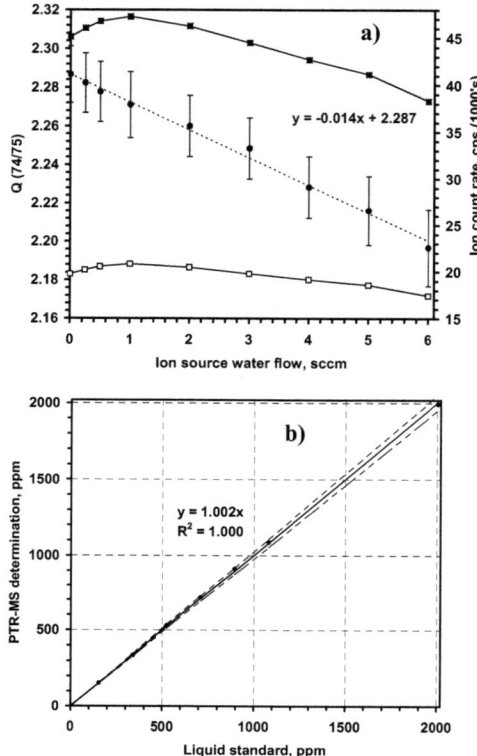

Figure 1. a) Q, the m/z 74 to m/z 75 ICR ratio for the headspace vapour of 2004 ppm deuterated water as ion source water flow is varied between 0 and 6 sccm. Also shown are the ion count rates for m/z 74 (closed squares) and m/z 75 (open squares) for the same range; b) PTR-MS determination of deuterium abundance in seven standard reference mixtures, corrected for ion-source water bleed.

Figure 2) presents PTR-MS determination of liquid-phase deuterium in human breath over a ~3 h period. Results are given for a 'control' subject (closed circles) and 'test' subject, both before and after ingestion of ~33 g (30mL) of pure D_2O. Data are obtained by determining the ICR ratio Q for consecutive measurements and applying the ion-source bleed correction expression described above. Liquid phase (i.e. body water) deuterium abundance is subsequently determined by applying a HDO partitioning coefficient of 0.937, which corresponds to a body temperature of $37°C$.

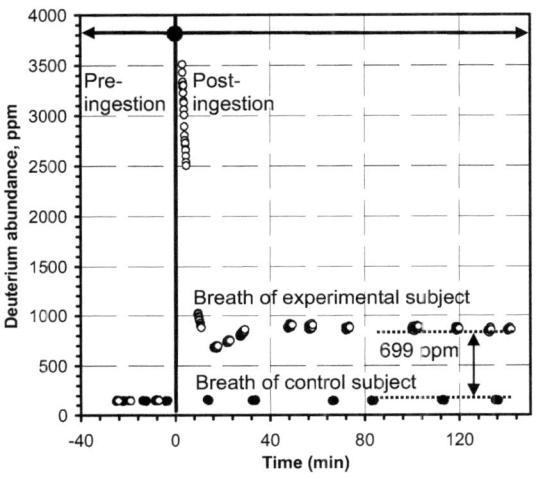

Figure 2. Deuterium abundance (ppm) in the breath of test and control subjects before and after consumption of 30mL D_2O and H_2O respectively.

Following consumption of D_2O, the deuterium content of the breath of the test subject increases rapidly, due to oral contamination. This contamination soon drops as the mouth is cleared by saliva and evaporation. A second maximum in deuterium concentration is observed between 30 and 40 min post consumption, corresponding to a flow of HDO from the upper intestine region into the blood stream[2]. This decreases to reach equilibrium after approximately 2 h, representing near total mixing of HDO throughout the body. Total body water (TBW) is determined from these equilibrium measurements using the simple method described by Spanel and Smith (2000), which compares the change in concentration of breath deuterium with the amount of D_2O consumed, as shown in Equation 2;

$$TBW = V(D_2O) / \Delta(D_{LIQ}) \qquad (2)$$

For the example given here, $V(D_2O)$ is 0.030 L and $\Delta(D_{LIQ})$ is 699 ppm, giving TBW volume as 42.9 L. The body mass of the test subject was 81 kg so we can easily determine that the test subject's body water is 53% by mass.

4. References

1. Spanel, P. and Smith, D. (2000) *J Am. Soc. Mass Spectrom.* **11**, 866-875
2. Smith, D. and Spanel, P. (2001) *Rapid Commun Mass Spectrom.* **15**, 25-32
3. Davies, S., Spanel, P. and Smith, D. (2001) *Physiol. Meas.* **22**, 651-659

Development of Proton Transfer Ion Trap Mass Spectrometry (PIT-MS): Identification of volatile organic compounds (VOCs) and inter-comparison with GC-MS

Carsten Warneke[1,2], Joost de Gouw[1,2], Edward R. Lovejoy[1]

[1]*Aeronomy Laboratory, National Oceanic and Atmospheric Administration, 325 Broadway, Boulder, CO*

[2]*Also with: Cooperative Institute for Research in Environmental Sciences, University of Colorado, Boulder, CO*

ABSTRACT

We present a newly developed instrument that utilizes Proton Transfer Ion Trap-Mass Spectrometry (PIT-MS) for on-line trace gas analysis of volatile organic compounds (VOCs). The instrument is based on the principle of PTR-MS (proton-transfer-reaction mass spectrometry), but as opposed to a quadrupole mass filter in a PTR-MS, the PIT-MS instrument uses an ion trap mass spectrometer, which has the following advantages: (1) the ability to acquire a full mass spectrum in the same time as one mass with a quadrupole, and (2) extended analytical capabilities of identifying VOCs by performing collision-induced dissociation (CID) and ion molecule reactions in the ion trap. The instrument described has a detection limit of <0.3 pbbv (S/N=3) for 1-min measurements for most tested VOCs. The PIT-MS was used for ambient air measurements during the NEAQS2004 campaign In New England onboard a ship, and agreed well with a GC-MS (gas chromatograph-mass spectrometer). Automated CID measurements on m/z 59 during the campaign are used to determine the contributions of acetone and propanal to the measured signal; both are detected at m/z 59 and thus indistinguishable in PTR-MS. It is determined that m/z 59 is mainly composed of acetone in agreement with the GC-MS.

1. Introduction

In a PTR-MS measurement, only m/z of the product ions can be determined, which is a valuable but certainly not a unique indicator of the identity of trace gases. It is clear that different isomers of the same mass cannot be resolved in this manner. The interpretation of the mass spectra is further complicated by the fragmentation of product ions and the formation of cluster ions, which may lead to additional mass overlap. Here we present an instrument similar to a PTR-MS that uses an ion trap mass spectrometer for the detection of ions. The extended analytical capabilities of the ion trap allow VOCs to be resolved that are normally indistinguishable in PTR-MS (Prazeller et al., 2003; Warneke et al., 2004). Furthermore, an ion trap can analyze a range of masses of several 100 amu almost simultaneously whereas the quadrupole mass filter transmits only ions of one mass at a time. The instrument design is described and performance tests are presented. Ambient atmospheric measurements of a number of different VOCs are used to demonstrate the instruments detection limit and the feasibility of on-line VOC measurements at low mixing ratios. The results are compared to GC-MS measurements. In addition, collision-induced dissociation (CID) measurements for ambient trace gases are used to demonstrate the capability of the PIT-MS instrument to distinguish between different VOCs of the same m/z.

2. Results and Discussion

2.1. Instrument Setup

The PIT-MS instrument shown in Figure 1 consists of four parts: (1) an all Teflon gas inlet and handling system, (2) an ion source for the production of $[H_3O]^+$, (3) a drift tube reaction chamber, and (4) the detection system with the ion trap mass spectrometer. The ion trap, used in the PIT-MS instrument, has been described in detail by Lovejoy and Wilson (1998). The cylindrical ion trap has an internal radius of 1 cm and stretched end cap geometry. The electronics and data acquisition system is custom built and the software gives the ability to choose the trapping time, to perform fast mass scans of all trapped ions, to selectively trap specific ions, do CID on selected ions, and to carry out ion-molecule reactions in the ion trap by adding reactant gases to the He buffer in the trap chamber.

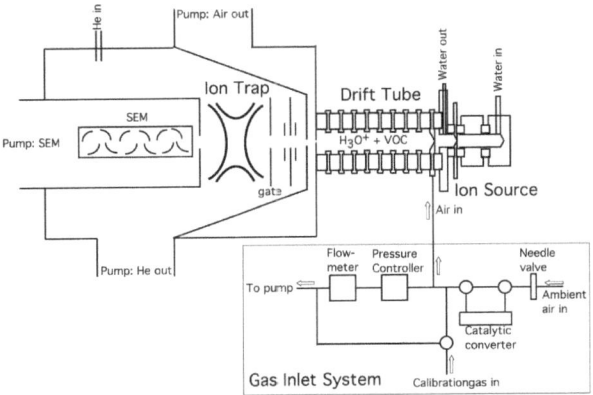

Figure 1: Schematic drawing of the PIT-MS instrument.

2.2. Ambient Air Measurements: Detection Limit and Inter-comparison with GC-MS

Organic compounds were measured by PIT-MS onboard the NOAA research ship Ron Brown during the New England Air Quality Study (NEAQS2004). The results obtained by PIT-MS during NEAQS were compared with GC-MS measurements of (oxygenated) hydrocarbons. Detection limits of PIT-MS are estimated as three times the standard deviation in the background measurements (Warneke et al., 2001), and are given in Table 1 for 1-min integration periods. For comparison, the detection limits for a PTR-MS determined during the previous NEAQS2002 campaign are also given in Table 1. The PTR-MS is still more sensitive than the PIT-MS at this point, but the latter is subject to continued improvement.

Compound	m/z [amu]	Det. Lim. [pptv]	det. lim. PTR-MS[pptv] (de Gouw et al., 2003)	Slope Inter-comparison	r^2 Inter-comparison
Methanol	33	300	250	0.75	0.96
Acetonitrile	42	40	34	0.82	0.80
Acetaldehyde	45	300	220	0.95	0.85
Acetone	59	150	84	1.00	0.97
Benzene	79	90	46	2.35	0.75
Toluene	93	50	33	1.20	0.96

Table 1: The best estimate of the sensitivity and detection limit (S/N=3) of the PIT-MS instrument for some VOCs and the slope and r^2 of the inter-comparison with the GC-MS.

2.3. Collision Induced Dissociation

The ion trap has extended analytical capabilities compared to the quadrupole mass spectrometer used in conventional PTR-MS. Using the PIT-MS instrument CID was performed every 3h on m/z 59 (sum of acetone and propanal) during NEAQS2004. Ions at m/z 59 were isolated in the ion trap and subsequently excited and fragmented using a filtered noise field (FNF) at increasing amplitudes. Figure 2a shows the time series of m/z 59 and Figure 2b shows the relative amount of acetone determined with the GC-MS. One typical CID measurement of m/z 59 and pure acetone and propanal CIDs are shown in Figure 2d-e. The ratio of (m/z 31)/(m/z 41), averaged over the FNF amplitudes 0.12V, 0.14V, and 0.16V, is plotted versus time in Figure 2c. For almost the entire measurement period the measured (m/z 31)/(m/z/41) ratio is very close to that of acetone. This shows that the signal measured on m/z 59 can be attributed mainly to acetone with a small contribution of propanal. This is confirmed by the GC-MS measurements.

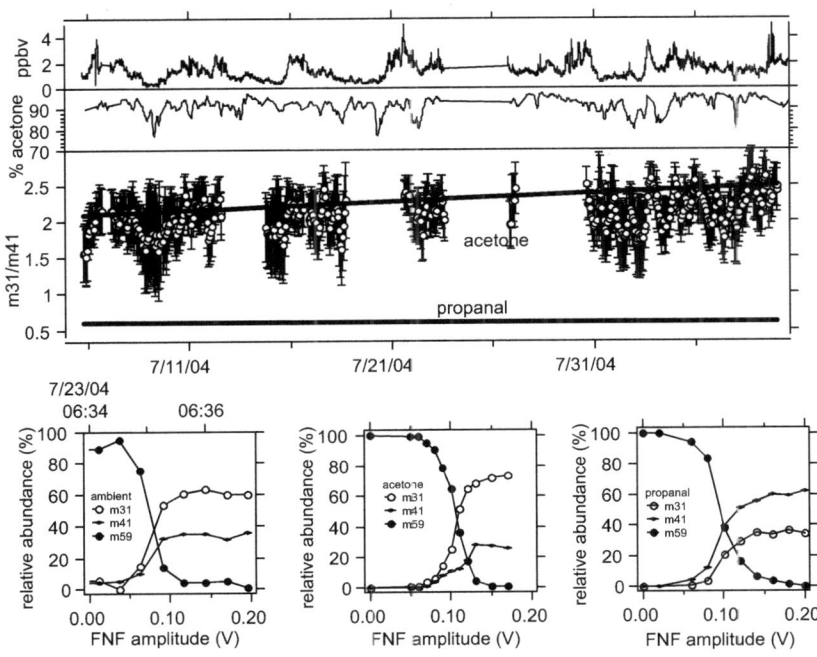

Figure 2: a) A time series of acetone mixing ratio. b) relative abundance of acetone determined by GC-MS. c) Ratio of (m/z 31)/(m/z 41) at an FNF amplitude of 0.14V versus time. The solid lines show the ratios for pure acetone and propanal. d) One single CID measurement of m/z 59 of ambient air. e) pure acetone. f) pure propanal.

References:

Lovejoy, E. R.; Wilson, R. R. *Journal of Phys. Chem. A* **1998**, *102*, 2309-2315.

Warneke, C.; van der Veen, C.; Luxembourg, S. L.; de Gouw, J. A.; Kok, A. *International Journal of Mass Spectrometry* **2001**, *207*, 167-182.

Prazeller, P.; Palmer, P. T.; Boscaini, E.; Jobson, T.; Alexander, M. *Rapid Communication in Mass Spectrometry* **2003**, *17*, 1593-1599

Warneke, C.; Rosen, S.; Lovejoy, E. R.; de Gouw, J. A.; Fall, R. *Rapid Communication in Mass Spectrometry* **2004**, *18*, 133-134.

de Gouw, J. A.; Goldan, P. D.; Warneke, C.; Kuster, W. C.; Roberts, J. M.; Marchewka, M.; Bertman, S. B.; Pszenny, A. A. P.; Keene, W. C. *J. Geophys. Res.* **2003**, *108*, 4682, 4610.1029/2003JD003863

Real-time VOC Monitoring Using PTR-MS Emitted from Hazardous Waste Incinerator Flue Gas

Akio Shimono[1] and Makoto Naganuma[1]

[1]Sanyu Plant Service Co., Ltd., 1-8-21 Hashmotodai, Sagamihara-shi, Kanagawa-ken, 229-1132 Japan (a.shimono@g-sanyu.co.jp)

ABSTRACT

Volatile Organic Compounds (VOC) emissions from flue gases were measured on-line at several incinerators in service using a proton transfer mass spectrometer (PTR-MS). The most abundant species in incinerator flue gases was m/z 45 considered mainly from acetaldehyde. Toluene was observed at relatively higher concentration among alkylbenzenes. The behavior of benzene was specific compared with other alkylbenzenes. The time trend of monochlorophenols were strongly correlated with spikes of benzene, while monochlorobenzenes, di- and tri- species of chlorobenzenes and chlorophenols did not keep track of the benzene variation. The on-line application of a PTR-MS on the process analysis in incinerators was considered to be effective for a developing better control technology of emissions from hazardous waste incinerators. Total emission amount of VOC species other than alkanes, ethylene and acetylene were roughly estimated and compared with the former study of VOC emission factor per automobile.

1. Introduction

Dioxins emission from municipal and hazardous waste incinerators are one of the largest public concerns about the environment. Though much efforts and measures to reduce dioxins emission have been taken, the complete understandings of their formation mechanisms for individual combustibles is still unreachable. Especially, for hazardous waste incinerators, the incineration conditions are always varies depending on types, chemical compositions, and physical forms of wastes. One small bottle of chlorinated organic reagent may violate the established condition in an incinerator. Though the on-line monitoring of some dioxin surrogates or indicators such as chlorophenols was successfully demonstrated [1], much effort will be needed to estimate which contributions is larger, the de-novo synthesis or the precursor mechanism [2] in the overall processes occurred in a real-world incinerator. While many evaluations of VOC emission have been done including fugitive emission from chemical plants, automobile exhausts and indoor environment, the emission factor for incinerators in the emission inventory is questionable. Some works [3] showed the existence of a large number of volatile organic species in an incinerator flue gas. However the transient behavior of most of VOC species are remained unknown.

PTR-MS attracts attention at the practical side. Namely the easier operation compared to other instrument systems employed REMPI or single-photon VUV ionization is ideal for incinerator applications as well as the linearity of measurement, high sensitivity, and high time-resolution. Moreover, as far as the assumption of a pseudo-first order reaction stands, we could approximate the mixing ratio level of an unidentified VOC species. This may be the first attempt to apply PTR-MS to on-line measurements of various VOC species in incinerator flue gases.

2. Experimental

The mass spectra for the representative species of interest were collected by using a high sensitivity version of PTR-MS. This preliminary study elucidated that the monitoring of chlorobenzenes and chlorophenols as dioxin surrogates was feasible at protonated molecular ion peaks considering the characteristic chlorine isotope patterns. We confirmed that the linearity of measurement by using standard gases containing some alkylbenzenes and short-chain alkenes in the laboratory. However the standard gas calibrations were not employed in the on-line measurements.

The VOC species existing in the incinerator flue gases were measured at four incineration sites with four rotary kiln incinerators, two fixed bed incinerators, and one liquid-injection incinerator in service. Except 1 incinerator site, other sites employed wet flue gas cleaning because bag filters were not efficient in high chlorine load often experienced in hazardous waste incinerators. Various types of hazardous wastes were incinerated in these sites. This presentation shows the typical cases observed in a rotary kiln having followed by a secondary combustion chamber, a waste heat boiler, a quench tower, a wet scrubber, and a wet electrostatic precipitator, which was certified as a chlorofluorocarbon (CFC) destruction facility depicted in Figure 1. Discarded CFCs or spent chloroform solvent are constantly incinerated at feeding of 20 kg/h which is 2.3 % of the incineration capacity. The total loading of halogen may frequently exceed 5 % summing up other feedings of halogenated wastes.

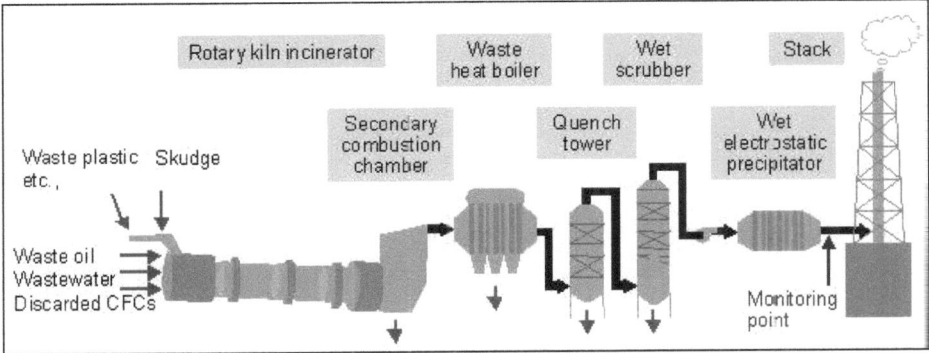

boiler) and after the flue gas cleaning system (at the inlet of the stack) were sampled in 3 L Tedora bags at the same incineration condition, both directly and by removing water vapor passing through a silica gel column. The measured mass spectra of collected samples were comparable between before and after the flue gas cleaning. The effect of water removal was also considered. Then we selected the on-line monitoring point after the flue gas cleaning system. The gas sampling setup included an auto-trap for condensed water, a heating of a line, and an empty impinger for removing large particles, and without any further pretreatment of gases. Masses up to 60 were scanned versus time, which were abundant or originated from some important species in the full spectra of measured gases. Temperature at the outlet of the rotary kiln and the secondary combustion chamber as well as CO, O_2, humidity, flow velocity, and temperature in the stack gas were simultaneously monitored. Temperature in stack gas when the waste heat boiler was operated and halted were around 62 C and 86 C, respectively And water vapor was saturated because of wet processes.

3. Results and Discussion

An example of time trend of some masses in the flue gases on a certain day is shown in Figure 2, where the time series of temperature at the outlet of rotary kiln and at the outlet of secondary combustion furnace, and CO concentration in the stack gas are also arranged using the same time axis. The waste heat boiler was operated as usual and chloroform solvent was burnt on that day enough to provide a large number of chlorine atoms for the formation or the occurrence of dioxin related species. The mixing ratio of acetaldehyde (m/z 45) was one to several orders of magnitude higher than other species and was maintained constant. Acetic acid possibly interfered with ethyl acetate (m/z 61) was also kept constant. The behavior of acetone + propanal (m/z 59) was different from other oxygenated species. Toluene (m/z 93) is the most abundant species among alkylbenzenes. The spike of benzene (m/z 79) shortly after the CO peak was observed, whereas toluene, C2-benzenes (m/z 107), and C3-benzenes (m/z 121) were not affected by this event. This spike exactly corresponded to the failure of waste feeding control to the incinerator. An isocyanate waste reacted violently with polyols probably existing in some other waste at that moment. The exothermic reaction caused a localized temperature rise inside the incinerator and a deficiency of oxygen to fuel ratio. No waste containing benzene itself was fed at that moment. The evidence of a rapid formation of benzene in an elevated temperature condition may indicate the dominance of the cyclization reaction through the excited state of acetylene in the flame zone of the incinerator. Acetonitrile (m/z 42) increased with some delay after the event.

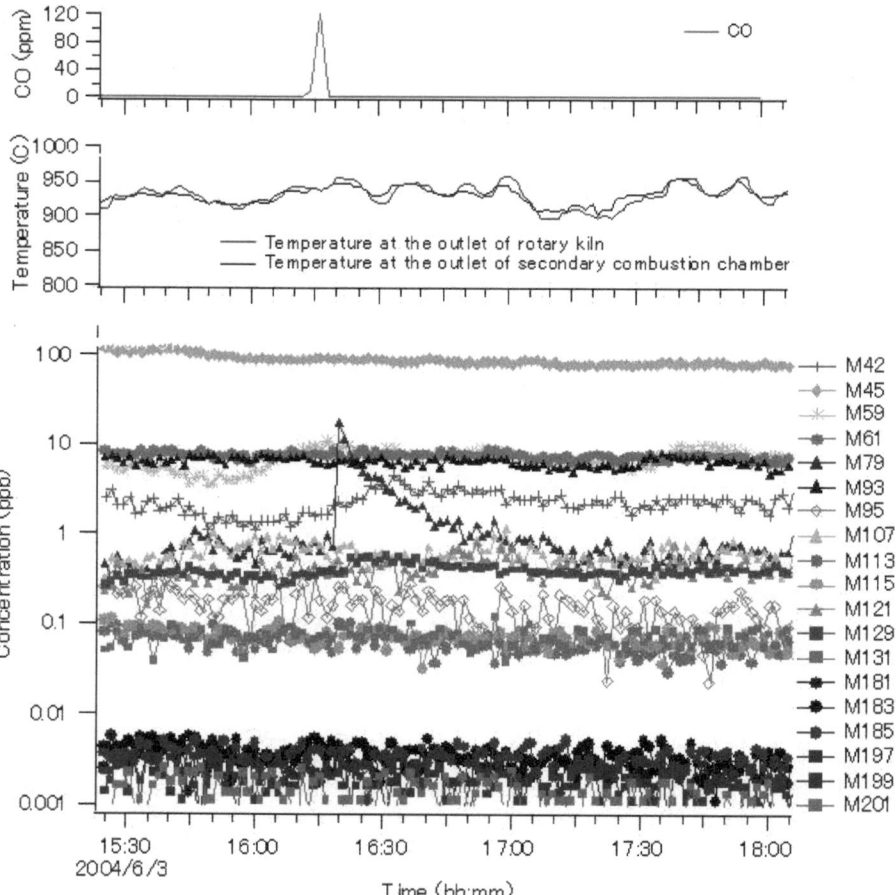

considered as the so-called precursor mechanism in the incinerator and the other may be the formation in the post-incinerator zone. The latter is also called as the memory effect which may produce a less variation of species relaxed by a larger time constant reflected the residence of soot particles in the flue system. Similar phenomena of a benzene spike followed by some chlorophenols increase was also observed at other incineration sites. Therefore benzene and chlorobenzene are considered as the most important indicators of reaction processes occurred in an incinerator. The level of monochloropehenols was lowered to 30 to

70 ppt on the day when the waste heat boiler was not operated. A waste heat boiler has been elucidated to offer a suitable temperature condition for the catalytic formation of dioxins and related species on accumulated fly ash. This result strongly supports the concept that the larger part of monochlorophenols was contributed by the de-novo synthesis in that incinerator. From a first approximation was employed using the mass spectra data taken on the day when the waste heat boiler was not operated. The total mixing ratio of measured VOC species is 213 ppb, half of which is the contribution by acetaldehyde. This value is converted to 0.372 mg/m^3 and results in 6.72 g/h by multiplying the measured gas flow rate. The VOC emission factor of 48.6 g/h for an ordinary diesel truck is calculated from the statistic data of 1.621 g/km in the speed region between 20 and 40 km/h by Ministry of Economy, Trade, Industry in 2001, assuming that a truck is drive at 30 km in a typical urban traffic condition. Taking the lack of information on alkanes, ethylene and acetylene by PTR-MS into consideration, whereas the above statistic data was based on the sampling analysis by GC-FID which may distort the mass loading of oxygenated species, the approximated total VOC emission per hour from a relatively well controlled incinerator is considered smaller than the value for an ordinary diesel truck.

4. Conclusions

The application of PTR-MS to on-line monitoring of incinerator flue gas has demonstrated and proven to be a powerful tool for process studies occurred in incinerators. The further investigations will be required for ensuring the accuracy of quantification by using standard gas calibrations of the representative species and comparing with a series of sampling analysis.

Future work will include the comprehensive analysis of both the gas phase and the particle phase coupled with a novel aerosol mass spectrometer. Our whole picture is to understand the processes actually occurred in incinerators with kinetic studies and to establish a better process control technology for further minimizing the VOC emissions from incinerators.

References

[1] Zimmermann, R., Hafner K., Dorfner R. Blumenstock R., and Kettrup A., On-line laser mass spectrometry for analysis of combustion processes: PCDD/F surrogates in waste incineration flue gases, *Organohalogen Compounds*, **2001**, 54, 368-373

[2] Tuppurainen, K., Halonen, I., Ruokojärvi, P., Tarhanen, J. and Ruuskanen, J., Formation of PCDDs and PCDFs in municipal waste incineration and its inhibition mechanisms: a review. *Chemospehere*, **1998**, 36, 1493-1511

[3] Jay K. and Steiglitz, L. Identification and quantification of volatile organic components in emissions of waste incineration plants, *Chemosphere*, **1995**, 30, 1249-1260

PTR-MS measurements in indoor environments

Armin Wisthaler[1], Charles J. Weschler[2], David P. Wyon[2], Pawel Wargocki[2], Armin Hansel[1], and Tilmann D. Märk[1]

[1] *Institut für Ionenphysik, Leopold-Franzens-Universität Innsbruck, Innsbruck, Austria*
armin.wisthaler@uibk.ac.at

[2] *International Centre for Indoor Environment and Energy, Technical University of Denmark, Lyngby, Denmark*

Indoor air research is receiving growing attention, reflecting the fact that in developed parts of the world people spend about 90% of their lives indoors. Among the indoor environments receiving attention have been those in office buildings and aircraft cabins. Primary emissions from building materials and furnishing, and secondary emissions generated from the reactions of ozone with gas-phase compounds or indoor surfaces, can degrade perceived air quality in office buildings and may contribute to Sick Building Syndrome (SBS) symptoms and ill health. Cabin air quality may induce perceived discomfort, respiratory ailments and other health problems in cabin occupants. In an attempt to investigate the role of volatile organic compounds (VOCs) in this context, PTR-MS was used as an analytical technique to monitor VOC levels in different indoor environments with different pollution sources present and with different air purification technologies in use. Sensory analysis was carried out for some of the studies to provide a measure of the human response to certain indoor environments.

The investigated scenarios included

- an office with different pollution sources (carpet, linoleum, chipboard) equipped with an air purification unit
- an office with typical indoor ozone levels and a limonene source (frequent component of air fresheners)
- a simulated aircraft cabin with ozone levels similar to those that have been observed during commercial air travel

The results indicate that the air purification unit can be efficiently used to reduce VOC loads in indoor environments and to improve air quality as perceived by humans. Ozone was found to be an important precursor for indoor VOCs. Ozone-initiated gas-phase chemistry (e.g. ozone-limonene reactions) formed a large spectrum of products including short lived, highly reactive compounds as well as secondary organic aerosols. In addition, a series of odour-active aldehydes was observed to be formed from ozone-fabric interactions in the simulated aircraft cabin. Secondarily generated species (both oxidized VOCs, and secondary organic aerosols) may be of greater concern to human health than exposure to ozone itself.

Selected topics of the conducted research will be presented.

A mobile FTICR for real time gas analysis

Joël Lemaire, Gérard Mauclaire, Pierre Boissel, Hélène Mestdagh and Michel Héninger

Laboratoire de Chimie Physique, UMR 8000 CNRS-Université Paris 11 Bâtiment 350, Université de paris Sud, France.

ABSTRACT

A small FTICR mass spectrometer allowing real times gas analysis with a high mass resolution has been developed in Orsay. The magnetic field is produced by a magnetic assembly based on the use of permanent magnet elements.

This apparatus is well adapted to the mass separation of small molecules that cannot be resolved by lower resolution instruments such as quadrupoles.

In the present apparatus proton transfer ionization can be realized in the ICR cell which is then both the reactor and the mass spectrometer. The sensitivity is limited to the ppm range but benefits of the high mass resolution of FTICR. A new design allowing the coupling with an external ion source will be presented. In this new configuration the proton transfer reaction will occur in an external reactor, giving access to high sensitivity.

1. Introduction

FTICR mass spectrometry is well known for its top performances concerning mass resolution and mass accuracy. However FTICR based on the use of supraconductor magnets are heavy, expensive and limited to use inside the laboratory.

With the use of a permanent magnet the size, weight (and cost) of the apparatus can be reduced tremendously. Of course the performances are more modest but they make such an instrument interesting for many applications.

Our first instrument, nicknamed MICRA (standing for Mobile ICR Analyzer) is used for infrared spectroscopic characterization of mass selected ions at the IR Free Electron Laser facility CLIO in Orsay. This application takes advantage of the design of the ICR cell which has a very open optical access.

We will focus here on the possible environmental applications of such an instrument and on how the formation of ions by proton transfer reaction can be coupled with FTICR mass spectrometry.

2. A mobile FTICR

The strength and homogeneity of the magnetic field are the most crucial parameters in a FTICR experiment. The seek for utmost performances, driven by the applications in the field of proteomics has pushed FTICR instrument development toward very big machines based on supraconductor magnets with fields up to 12 Tesla being now commercially available.

The apparatus developed in Orsay has a moderate 1.25 Tesla nominal magnetic field and a field homogeneity close to 10^{-3} in the 8 cm^3 cell volume. Its weight is about 200 kg and its dimensions are $0.6 \times 1.2 \times 0.8$ m.

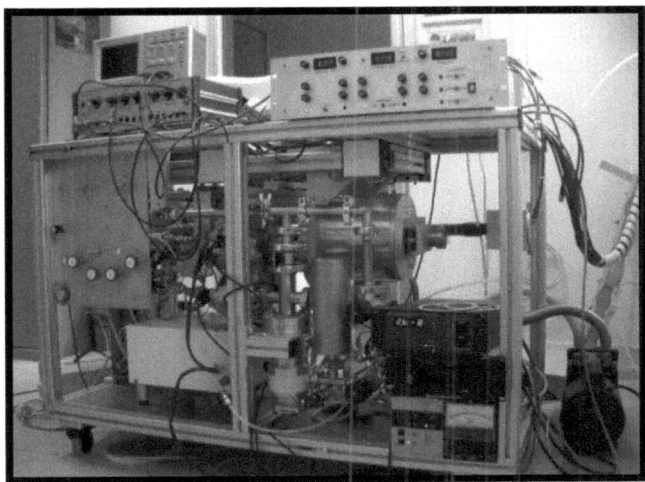

Fig1. View of our first prototype, 'MICRA' ('Mobile ICR Analyzer')

The magnet

The magnet is a magnetic assembly made of two so called 'Halbach cylinders'. One cylinder is composed of 8 magnetic segments, the magnetization direction being tilted by 90 degrees from one segment to the next. These magnetized elements are made of Neodymium-Iron-Boron (a material with a high magnetic energy).

The distance between the two Halbach cylinders is set so as to optimize the field homogeneity in the z direction (along the cylinders revolution axis).

This design produces a very homogenous magnetic field in the bore of the cylinder and a very limited stray field outside.

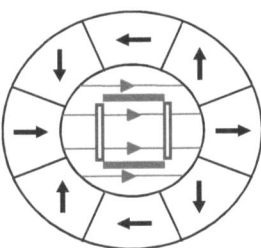

Fig2. Schematic of the Halbach cylinder showing the magnetization direction of the 8 segments.

Available modes of ion formation

Because of the magnetic field configuration (the magnetic field being perpendicular to the cylinder axis) the ions must be produced either inside the ICR cell or in its close vicinity.

Electron impact ionization and chemical ionization are easily implemented. In the electron impact mode a collimated pulsed electron beam ionizes the gas inside the cell. Gases are introduced in a pulsed way through the combination of a leak valve and a three way pulsed valve directing the gas flow either in the mass spectrometer main chamber or toward the gas inlet turbomolecular pump.

Chemical ionization is achieved by producing the precursor ion in a first gas pulse, eventually followed by an ejection pulse to ensure that only one type of ion is kept in the cell. The precursor ion is then allowed to react with a second gas pulse. The extent of reaction is controlled by tuning the pulse duration and intensity.

Metal ion precursors can be formed by focussing the 355nm output of a NdYAG laser on a metal target located in front of a 5mm diameter hole in one of the trapping. The same laser can also be used at smaller intensities to produce ions of biological interest by MALDI.

Main characteristics

This device benefits of the main advantages of the FTICR mass spectrometry technique :

-robustness

-simultaneous detection of the full mass spectra

-high mass resolution and accuracy

-it is a trapping technique, allowing for complex sequences with several reaction and selection steps

-if necessary (when mass spectrometry alone is not enough) ions can be further characterized using chemical or physical techniques (by their reactivity, by photodissociation) so that even isomers can be distinguished.

2. High mass resolution with MIMs introduction

Coupling the FTICR with a membrane inlet introduction allows PTRMS analysis of aqueous solutions. We show here that high resolution detection is possible in this configuration with an example where two species of the same nominal mass are present as traces in water. The membrane interface is made of a 3mm outer diameter silicone tube in which the solution to analyze is flowing. The silicone tube is enclosed in a stainless steel tube connected to the mass spectrometer inlet. Methyl acetate (CH_3COOCH_3) and diethylether ($C_2H_5OC_2H_5$) present at a few ppm concentration in water are detected using proton transfer reaction from H_3O^+ ions in the ICR cell. The protonated ions of diethylether (m/z=74.907) and methyl acetate (m/z=74.939) both have the same nominal mass but are clearly resolved on the mass spectra shown on fig. 3.

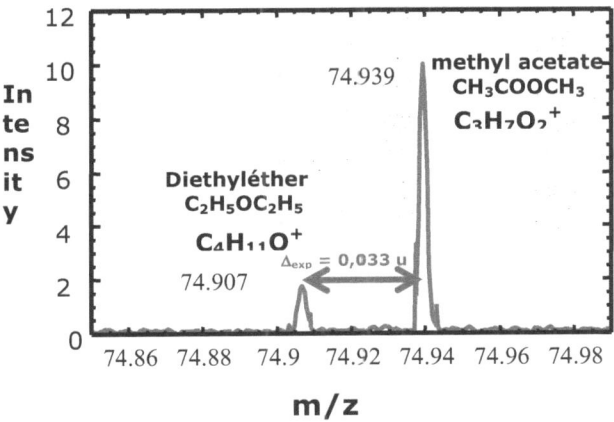

Fig 3. Mass spectra resulting from the membrane inlet introduction of diethyl ether and methyl acetate both present at trace concentrations in water.

3. New design

A new design for the permanent magnet assembly is under realization. In this new design the magnetic field produced by a cylindrical magnetic assembly is collinear with the cylinder revolution axis thus making possible the injection into the mass spectrometer of ions produced in an external source. This will enable the coupling with different ion sources and in particular with a proton transfer external reactor. With differential pumping between the external source and the vacuum housing containing the cell it will be possible to maintain the very low vacuum necessary for FTICR detection while reacting the H_3O^+ precursor with air at pressures up to 1 torr.

The expected homogeneity (from the magnetic simulation calculations) with this new magnet configuration should be better than with the the system based on two Halbach cylinders and the space available for the cell will be somewhat increased with a bore diameter of 6 cm.

The first tests of this new configuration will take place in the beginning of 2005.

4. Acknowledgment

We aknowledge the CNRS and of the University of Paris-Sud 11 of their financial support for the construction of the new instrument (ACI and BQR). We also acknowledge CLIO for the acquisition of the ablation laser, through the financial support of the Laser center POLA.

5. References

(1) MICRA : A compact permanent magnet FTICR mass spectrometer.
G. Mauclaire, J. Lemaire, P. Boissel , G. Bellec, M. Heninger
Eur. J. Mass Spectrom. 10, (2), 155-162, (2004).
(2) Gas Phase IR photodissociation spectroscopy using an FTICR ion trap coupled to a free electron laser

J. Lemaire, P. Boissel, M. Heninger, G. Mauclaire, G. Bellec, H. Mestdagh, A. Simon, S. Le Caer, J.M. Ortega, F. Glotin and P. Maitre   Physical Review Letters 89(27) 273002 (2002)

(3) Structural characterization of selectively prepared cationic iron complexes bearing monodentate and bidentate ether ligands using infrared photodissociation spectroscopy
S. Le Caer, M. Heninger, J. Lemaire, P. Boissel, P. Maître, H. Mestdagh
Chem. Phys. Lett., 385, 273(2004)

(4) Infrared Multiphoton Dissociation Spectroscopy of Gas Phas Mass-Selected Hydrocarbon-Fe+ complexes
Aude Simon, William Jones, Jean-Michel Ortega, Pierre Boissel, Joël Lemaire, Philippe Maître
accepted in JACS

2.2 Food Technology

Individual Aroma Release & Perception

Peter Prazeller, Nicholas Antille, Santo Ali, Philippe Pollien, Laurence Mioche

*Nestlé Research Center, Vers-chez-les-Blanc, 1000-Lausanne 26, Switzerland,
(peter.prazeller@rdls.nestle.com)*

ABSTRACT

A large individual variation of in-vivo aroma release during eating exists. The question to
answer is to what extent these individual differences have an impact on perception. To assess
this 14 subjects (age 25–49, average 31 years) performed a sensory test while simultaneously
the in-mouth aroma release was measured. Samples consumed were 10ml aqueous ethyl
butyrate solutions in concentrations of 6, 6.5, 7.2, 8.7, 10.4,12.5 and 15 ppmV. A 3-AFC
(three alternative forced choice) test was conducted at the same time with in-vivo aroma
release analysis. Three ethyl butyrate solutions were presented to the panellists; two of them
were the reference at 6 ppmV, one the target at varying concentrations from 6.5 to 15 ppmV.
The subjects had to choose which of the samples was the stronger one. A correlation between
individual aroma release and aroma perception could be confirmed. Higher ability to
discriminate between samples of low concentration difference was found for those subjects
that showed a higher in-mouth release and a lower variability.

1. Introduction

During eating or drinking food aroma gets released into the oral cavity and transported with
the airflows associated with chewing, swallowing and breathing via the retronasal route to the
receptors in the olfactory epithelium in the nasal cavity. It is there where by selective binding
of odor molecules odor receptors are activated resulting in a cascade of signal transductions in
the olfactory neurons, the odor perception has its initial step. Therefore knowledge of the
aroma at this point is highly desirable. One way to assess this analytically is by measuring the
aroma composition in the exhaled air through the nose, because that is the same air that passes
over the receptors. This technique, often called nosespace analysis, is more and more used (1-
8) and represents a step forward in investigating the molecular basis of odor perception during
food consumption.

Strong differences in aroma release between subjects have been seen, which might be at least
partly the explanation of the individual differences seen in perception and consequently in
customer satisfaction. The objective of this study is to understand to what extent individual
differences of in-vivo aroma release during eating are the basis of individual aroma
perception. This is achieved by measuring the aroma release of simple model system of aroma
in water while simultaneously conducting sensory evaluation.

2. Materials & Methods

Sample Materials:

Samples consumed in this study were water solutions of ethyl butyrate (Sigma-Aldrich
Chemie GmbH, Germany, natural, FCC FEMA 2427), presented at room temperature in a
cup. Concentrations were 6, 6.5, 7.2, 8.7, 10.4,12.5 and 15 ppmV, the volume was 10ml.
Purity way checked by gas chromatography olfactometry (GC-O) analysis.

In-Vivo Aroma Release Analysis:

The most effective way to measure the release of aroma during eating is to monitor the breath-air as close as possible to the olfactory receptors in the nose. As for ethical and practical consideration it is difficult to sample the air directly in the close vicinity of the receptors, one has adopted the approach to collect the exhaled air at the nostril, breath-by-breath.

To provide comfort during experimentation, an ergonomic nosepiece allows panelists to breathe freely and comfortably. This is important to mimic a natural eating situation. The air exhaled through the nose is sampled via two glass-tubings. Each panellist has an individual nosepiece, tailor-made in order to smoothly and comfortably fit into the nostrils. The nosepiece is fixed on laboratory eyeglasses, and connected to the analytical instrument via flexible and heated tubing. The instrument used is a Proton-Transfer-Reaction Mass-Spectrometer (PTR-MS). It allows real time analysis of the aroma compounds as they are released during the eating process. A schematic of the set-up and the nosepiece are shown in Figure 1 below.

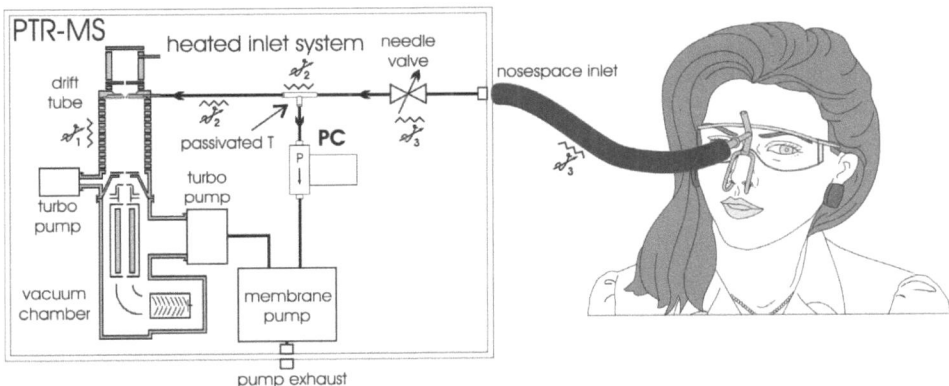

Figure 1: Schematic of the nosepiece for sampling breath-by-breath the air exhaled through the nose during eating and drinking. The tube is inactivated with inner quartz coating ("silcosteel®"-tube from RESTEK, Bad Homburg, Germany). These 80 ml/min are split into two fractions: 14 ml/min is introduced into the drift tube of the PTR-MS, and the remainder is released through a pressure controller and membrane pump into the laboratory air. All tubings are heated to 70°C to prevent condensations.

Sensory Evaluation:

14 panellists (age 25–49, average 31 years) participated in the study. The panel consisted of subjects who had experience with aroma release analysis and sensory evaluation as well as of people with little or no prior experience at all. In training sessions all were familiarized with nosespace analysis and the samples to be consumed, in the concentration range used in the test.

A 3-AFC (three alternative forced choice) test was conducted simultaneously with in-vivo aroma release analysis. Three ethyl butyrate solutions were presented to the panellists, two of them were the reference at 6 ppmV, one the target at varying concentrations from 6.5 to 15 ppmV. The subjects had to decide which of the samples was the stronger one. For every concentration pair 20 replicates were done.

Statistical analysis:

Most parameters appeared to be very heteroscedastic. This means that the data variability depends on data absolute intensities. Thus, the data needed to be transformed to make them more homoscedastic, which was achieved simply by a log-transformation. All statistical analyses were based on log-transformed data. These analyses consisted on the one hand of a descriptive analysis and on the other hand of more classical statistics, essentially analysis of variance (ANOVA) and Least Significant Difference (LSD).

3. Results and Discussion

As the nosespace signal intensities are time-dependent, it was necessary to extract a reduced list of parameters prior to performing effect analysis. The prospective character of this study led to a selection of most intuitive parameters. These are first the "swallow peak", the maximum compound concentration during the first breathing cycle after swallowing and secondly the "total area", the cumulated area of all breathing cycles for each sample. Included in this contribution are only data of the swallow peak.

Links between individual aroma release and how aroma is perceived could be confirmed. Higher ability to discriminate between samples of low concentration difference was found for subjects with higher aroma release in general. As an example Figure 2 shows the median concentrations for the 15 ppmV solution measured in the nosespace for different assessors plotted vs. the percentage of correct responses during all tests. A clear relationship could be found. People with higher release are better in distinguishing between samples.

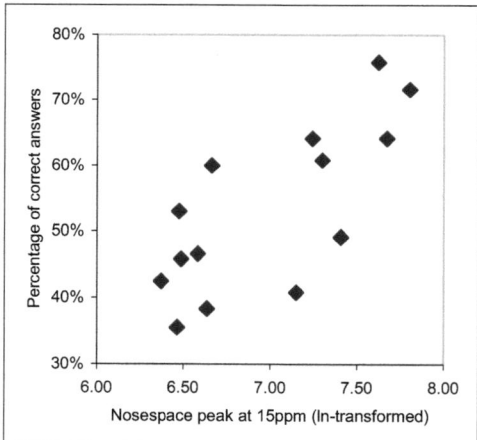 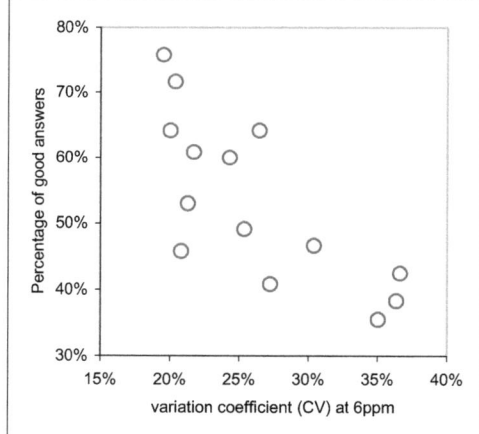

Figure 2: Correlation between aroma release concentration for a 15 ppmV solution (ln-transformed) and sensory performance for 14 subjects. Each marker represents median values of 20 replicates for one subject.

Figure 3: Correlation between variability (variation coefficient CV) of aroma release for the 6 ppmV reference solution and sensory performance for 14 subjects.

A second finding is that higher ability to discriminate between samples was found for people with the lowest variation in the release of the reference solution. As the task was to compare various concentrations to a reference of 6 ppmV, the more reproducible the aroma release was for this reference, the better subjects were in discriminating the stimuli. This is represented in Figure 3 by plotting the variation coefficient (CV) of the aroma release of 6 ppmV solution vs. the overall percentage of correct given answers. People with largest variation tend to be worst in fulfilling the task and vice versa.

As the panel consisted of people with varying degree of experience both in in-vivo aroma release analysis and in sensory evaluation the possibility of a correlation between experience and sensory performance was investigated but could not be confirmed. Nevertheless, an interesting experiment would be it to follow the development of subject's reproducibility in aroma release and in sensory evaluation during a panels training period.

4. Conclusion:

These results link individual aroma release to perception. It is well know that aroma perception is not only determined by the aroma concentration at the receptors, but also affected by a variety of factors like memory, mood, expectations, age or individual sensitivity. However, findings from this study strongly support the importance of the aroma concentration itself as it showed that even individual differences in aroma release have a large influence on the olfactory capabilities. Therefore individual differences observed in perception at least to a certain extent can be explained by in between subject variability of aroma release during consumption.

References

1. Soeting, W. J.; Heidema, J. A mass spectrometric method for measuring flavour concentration/time profiles in human breath. *Chem. Senses* **1988,** *13* (4), 607-617.
2. Taylor, A. H.; Linforth, R. S. T. Direct mass spectrometry of complex volatile and non-volatile flavour mixtures. *Int. J. Mass Spec.* **2003,** *223-224* (15 Jan 2003), 179-191.
3. Taylor, A. J. Release and Transport of Flavors *In Vivo*: Physicochemical, Physiological, and Perceptual Considerations. *Comprehensive Reviews in Food Science and Food Safety* **2003,** *1* (2), 45-57.
4. Hansson, A.; Giannouli, P.; van Ruth, S. The influence of gel strentgh on aroma release from pectin gels in a model mouth and in vivo, monitored with proton-transfer-reaction mass spectrometry. *Jou* **2003,** *51* (16), 4732-4740.
5. Roberts, D. D.; Pollien, P.; Antille, N.; Lindinger, C.; Yeretzian, C. Comparison of nosespace, headspace, and sensory intensity ratings for the evaluation of flavor absorption by fat. *J. Agric. Food Chem.* **2003,** *51* (12), 3636-3642.
6. Pionnier, E.; Chabanet, C.; Mioche, L.; Le Quere, J. L.; Salles, C. 1. In vivo aroma release during eating of a model cheese: Relationships with oral parameters. *J. Agric. Food Chem.* **2004,** *52* (3), 557-564.
7. van Ruth, S. M.; Roozen, J. P.; Cozijnsen, J. L. Volatile compounds of rehydrated french beans, bell peppers and leeks. Part 1. Flavour release in the mouth and in three mouth model systems. *Food Chem.* **1995,** *53,* 15-22.
8. Roberts, D. D.; Acree, T. E. Simulation of Retronasal Aroma Using a Modified Headspace Tewchnique: Investigating the Effects of Saliva, Temperature, Shearing, and Oil on Flavor Release. *J. Agric. Food Chem.* **1995,** *43,* 2179-2186.

Dynamics of Retronasal Aroma Perception during Consumption:
Cross-linking Diverse Analytical Tools for Elucidation of a Complex Process

Andrea Buettner

[1]*Deutsche Forschungsanstalt fuer Lebensmittelchemie, Lichtenbergstr. 4, D-85748 Garching, Germany, e-mail : Andrea.Buettner@Lrz.tum.de*

In the last years, research on retronasal aroma perception has been dominated by "nosespace"- and "mouthspace"-analyses involving different trapping or mass spectrometric techniques [1,2,3]. The focus laid on physical release phenonema such as partitioning of odorants between polar and non-polar phases and air. However, it is likely that temporal resolution and intensity of retronasal aroma perception during food consumption are highly influenced not only by food matrix composition but also by physiological factors. Plausible factors are oropharyngeal performances during mastication and swallowing [4], the adsorptive potency of odorants to oral mucosa [5] or the formation of adhesive coatings by food matrix constituents on oral and pharyngeal mucosa [4]. Also, salivary constituents can influence the concentrations and release patterns of odorants during consumption [1,6,7]

To draw a complete picture of consumption parameters and the dynamics of retronasal aroma perception, an array of analytical techniques needs to be applied, most preferably in combinations. Illustrated by the evaluation of wine, such a cross-linking approach will be demonstrated, combining online breath analysis via PTR-MS, together with EXOM (Exhaled Odorant Measurement)-, SOOM (Spit-Off Odorant Measurement)-, BOSS (Buccal Odor Screening System)-, and Saliva Assays into a comprehensive analytical tool [3,5,8]. The analytical data is complemented by time-resolved sensory analysis, as well as visualization of the processes using videofluoroscopy and real-time magnetic resonance imaging of the oropharyngeal areas of the human body.

References

1   Taylor, A.J. Crit. Rev. Food Sci. Nutr. 1996, 36, 765-784.
2   Hansson, A.; Giannouli, P.; van Ruth, S. J. Agric. Food Chem. 2003, 51, 4732-4740.
3   Buettner, A.; Schieberle, P. Food Sci. Technol. 2000, 33, 553-559.
4   Buettner, A.; Beer, A.; Hannig, C.; Settles, M. Chem. Senses, 2001, 26 (9), 1211-1219.
5   Buettner, A.; Schieberle, P. In: Flavor release (Roberts, D.D.; Taylor, A.J.; eds), ACS Symp. Ser. 763, 2000, pp. 87-98.
6   Buettner, A. J. Agric. Food Chem. 2002, 50 (11), 3283-3289.
7   Buettner, A. J. Agric. Food Chem. 2002, 50, 7105-7110
8   Buettner, A.; Welle, F. Flavour Fragr. J. 2004, 19, 505-514.

Characterization of Agricultural Processes in the Growth and Preservation of Fruits by PTR-MS VOC Monitoring

M. Vescovi[1], A. Weber[1], D. Barbon[1], A. Tonini[1], L. Fadanelli[2], M. Comai[2], G. Stoppa[3], R. Verucchi[1], A. Boschetti[1] and S. Iannotta[1].

[1]*Institute of Photonics and Nanotechnology, ITC Division, Via Sommarive 18, 38050 Povo di Trento – Italy – iannotta@itc.it*

[2]*Istituto Agrario di S. Michele all'Adige, 30010 S. Michele all'Adige, Trento Italy.*

[3]*Dipartimento di Informatica e Studi Aziendali, Università degli Studi di Trento, Via Inama 5- 38100 Trento Italy*

ABSTRACT

We report on the study of the effects of the altitude of the site of production and of the tree crop load on the VOCs emission of Golden Delicious apple. The sistematic measurements carried out by PTR-MS on two years, over four different sites of production and discriminating three different crop load on the tree, show interesting correlations. We demonstrate the ability of VOCs analyzed, during a preservations period of 7 months, of discriminating both site of production and crop load.

1. Introduction

The question of defining the quality of fruits and of apples in particular is a very difficult one. It is a common knowledge that it is relatable to flesh firmness, soluble solids, acidity and also to antioxidant substances and volatile organic compounds (VOCs). An increasing interest has been recently devoted to the ability of VOC detection to determine a non destructive way to quality assessment[1,2]. In this contribution we adopt such an approach to address the question of crop load that is a well known factor affecting fruit quality[3]. High leaves to fruit ratio is linked to firm flesh, high soluble solids and acidity content[4,5]. Some authors [6,7] did not find a clear relation between crop load and inside quality. While Poll et al.[8] and Werheim[9] found that low crop load induces more precursors for polyphenols biosynthesis, in other works[10,1] no relation was noticed between crop load and precursors of polyphenols. The aroma pattern is another important quality factor, which according to some authors is affected by crop load[8] while for others the connection is of scarce importance[12].The relationship between crop load and quality, already controversial at harvest, is even less known after storage. With the aim of better understanding this relationship, we monitored the VOCs emission using PTR-MS during storage of apple of Golden Delicious, produced by trees with different crop load and grown at various altitudes in Non valley, an important apple production area of Trentino (Italy).

2. Materials and Methods

In spring 2002 four adult Golden Delicious orchards were chosen at altitudes of 340 (Denno), 540 (Torra), 740 (Tres) and 940 (Romeno) m a.s.l. In each site, thirty homogeneous adult trees (spindelbush trained and grafted on M9 rootstock) were selected to form, according to the orchard architecture, three crop load levels of ten trees each from low to medium and high crop load. The three crop loads were achieved by hand thinning, aiming at three different fruit number per trunk section, according to the three defined crop load levels. Harvest time was decided on the ground of the ripening indices commonly used by the Organization of

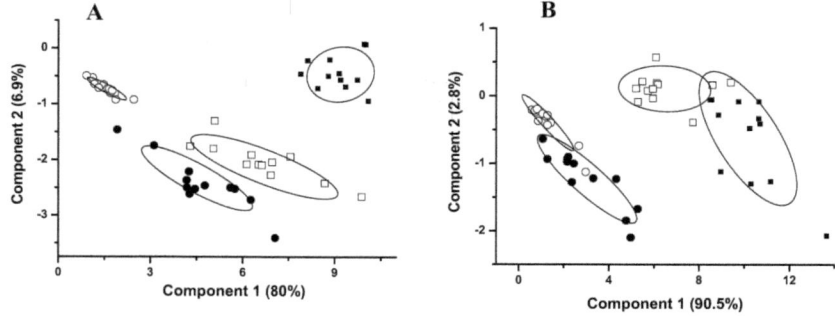

Figure 4 Relative Principal Components Analysis on VOCs data, normalized by weight. We have two period of measurement: February 2003 (A) and February 2004 (B). Apples come from different places: Denno (●), Torra (○), Tres (■), Romeno (□). The ellipses are draw at 68% of confidence level.

Producers in Trentino. At harvest, from each tree, a sample of 30 fruits within 70 to 80 mm in diameter was collected for storage and further analysis. Apple samples were stored in C.A. (1,5% O_2, 2,5% CO_2, 1 °C, 95% U.R.) for eight months, from September 2002 to May 2003. The campaign of experiments was reproduced using the same procedures in the year after (September 2003- May 2004).The study of VOCs was carried out on 5 intact fruits and

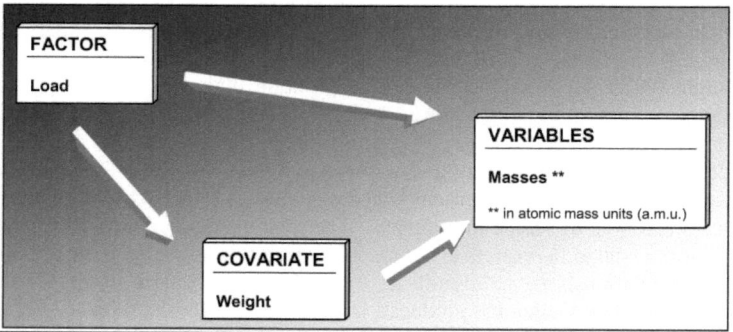

Figure 2 Direct Acyclic Graph (DAG): each node is a variables block. Each arrow represents a probabilistic dependence and can be interpreted as potential cause.

repeated for 4 times on different equivalent samples. The experiments considered apples after 5 and 8 months of storage. After extraction from the storage room the samples were kept for 7 days at temperature of 5 C° before the measurements in order to simulate the retail trade period. The headspace of the 5 apples for each test case was monitored by means of our Proton Transfer Reaction Mass Spectrometer (PTR-MS)[1, 2, 13]. Data were statistically analysed using the analysis of the relative principal components (R-PCA) where the term relative refers to the way the coefficient of variation are calculated[14].

3. Results and Discussion

Figure 1 compares the data obtained in the two different years of the study both referred to apples analyzed after 5 months of preservation. The R-PCA plots show in both cases a very significant discrimination of the apples coming from the different production sites positioned at the different altitudes from the sea level. This discrimination is consistently observed at all the different period of preservation considered. Even though the two years of production

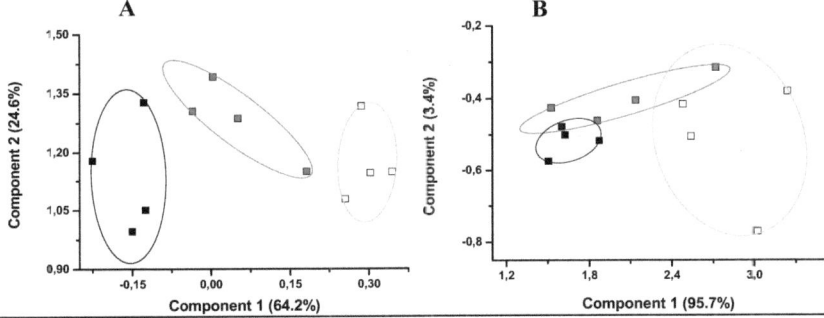

Figure 3 Relative Principal Components Analysis on masses selected with a statistical causality model. VOCs data refer to apples coming from the places of Tres at two different period: February 2003 (A) and May 2003 (B). Three level of carica are distinguished: high (■), medium (■), low (□). The ellipses are draw at 68% of confidence level.

considered have been quite different from the point of view of the climate the data show a very similar trend so that we believe they give a very strong indication that the VOC monitoring gives information about the effect of the altitude on the ripening and quality of production induced by the specific site.

In order to better discriminate the effects of the crop load at the different sites of production, we needed to analyze statistically which masses in the PTR-MS spectra were most sensitive to it. To this end we have developed a statistical approach[14] the basic scheme of which is reported in figure 2. It is basically a Direct Acyclic Graph where each node is variable bloc

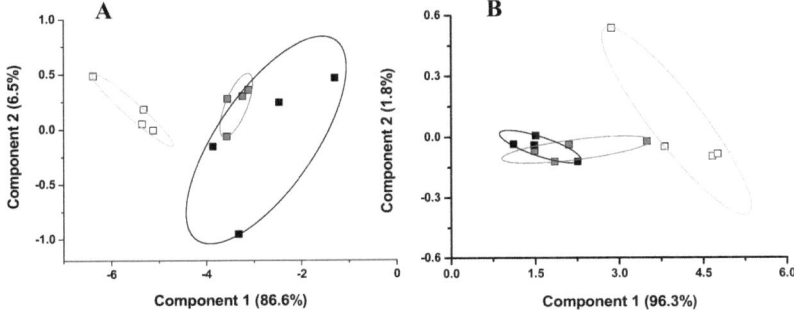

Figure 4 Relative Principal Components Analysis on masses selected with a statistical causality model. VOCs data refer to apples coming from the places of Denno at two different period: February 2003 (A) and May 2003 (B). Three level of carica are distinguished: high (■), medium (■), low (□). The ellipses are draw at 68% of confidence level.

and each arrow represents a probabilistic dependence that can be interpreted as a potential cause. The single variable bloc reflects the logical-temporal order on the basis of consequentiality of various phases of the experiments. Figure 3 and 4 shows the results of the analysis in terms of a R-PCA plot for two different site of production (Tres and Denno respectively). The major difference between the hight crop load and the lowest one is remarkably well discriminated in both cases. The intermediate load is instead overlapping one of the two other cases considered. It should be mentioned that the spread between the crop loads is relatively small (from about 4 to 10 fruits per cm^2 of the section of the tree trunk) so that the result obtained seem to be very interesting to assess the final quality of the products.

4. Acknowledgement

This work was financially supported by the Autonomous Province of Trento, Fondo Unico 2001, Project Qualiquant (Relationships between fruit crop load and quality in apple: an interdisciplinary approach) and by the "Convenzione PAT-CNR" Project: Analysis and researches on the agro-industrial system (Subproject3).

5. References

1. Biasioli, F., Boschetti, A., Toccoli, T., Jordan, A., Fadanelli, L., Lindinger, W. and Iannotta, S. Proton transfer reaction mass spectrometry : a new technique to assess post harvest quality of strawberries. Proc. 4th Int. Strawberry Symp. Acta Hort. **567** (2002), 739-742.

2. Boschetti A, Biasioli F, van Opbergen M, Warneke C, Jordan A, Holzinger R, Prazeller P, Karl T, Hansel A, Lindinger W and Iannotta S, 1999. PTR-MS real time monitoring of the emission of volatile organic compounds during post-harvest ageing of berry fruit. Postharv Biol Technol **17** (1999), 143-151.

3. Johnson D.S. The effect of flower and fruit thinning on the firmness of "Cox's Orange Pippin" apples at harvest and after storage. Journal of Horticultural Science **67**(1) (1992), 95-101.

4. Johnson D.S., Influence of time of flower and fruit thinning on the firmness of "Cox's Orange Pippin" apples at harvest and after storage. Journal of Horticultural Science **69** (2) (1994), 197-203.

5. Link H. Significance of flower and fruit thinning on fruit quality. Plant growth regulation 31: 17-26.

6. Ferree D.C., Cahoon G.A. 1987. Influence of leaf fruit ratios and nutrient sprays on fruiting, mineral elements, and carbohydrates of apple trees. J. Amer. Soc. Hort. Sci. **112**(3) (2000), 445-449.

7. Ferguson I.B., Watkins C.B. Crop load affects mineral concentrations and incidence of bitter pit in "Cox's Orange Pippin" apple fruit. J. Amer. Soc. Hort. Sci. **117**(3) (1992), 373-376.

8. Poll L., Rindom A., Toldam-Andersen T.B., Hansen P., Availability of assimilates and formation of aroma compounds in apples as affected by the fruit/leaf ratio. Physiologia Plantarum **97** (1996), 223-227.

9. Wertheim S.J., Developments in the chemical thinning of apple and pear. Plant Growth Regulat. **31** (2000), 85-100.

10. Awad M.A., De Jager A., Dekker M., Jongen W.M.F., Formation of flavonoids and chlorogenic acid in apples as affected by crop load. Scientia Horticolturae. Vol. **91** (2001), 227-237.

11. Stopar M., Bolcina U., Vanzo A., Vrhovsek U. 2002. Lower crop load for "Jonagold" apples (Malus x domestica Borkh.) increase polyphenols content and fruit quality. J. Agric. Food Chem., in press.

12. Mpelasoka B.S., Behboudian M.H., Production of aroma volatiles in response to deficit irrigation and to crop load in relation to fruit maturity for "Braeburn" apple. Postharvest Biology and Technology **24** (2002), 1-11.

13. Lindinger W, Hansel A and Jordan A, 1998. Proton-transfer-reaction mass spectrometry (PTR-MS), on line monitoring of volatile organic compounds at pptv levels. Chem. Soc. Rev **27** (1998), 347-354.

14. Barbon, D., Weber, A., Vescovi, M., Tonini, A., Boschetti, A., Iannotta, S., Fadanelli, L. and Stoppa, G.,. A statistical approach for the analysis of proton transfer reaction mass spectrometry (PTR-MS) data aimed at a qualification of fruits based on VOC emissions. Proc. 5th Int. Postharvest Symp. Verona, Italy 6-11 June (2004).

PTR-MS in agroindustrial applications: a methodological perspective

Franco Biasioli[1], Flavia Gasperi[1], Eugenio Aprea[1,2], Daniela Mott[1], Isabella Endrizzi[1], Valeria Framondino[1], Tilmann D. Märk[2]

[1]*Istituto Agrario San Michele all'Adige, Via E. Mach 2- 38010 San Michele all'Adige (TN)-Ital. Flavia.gasperi@ismaa.it*

[2]*Institut für Ionenphysik, Universität Innsbruck, Technikerstrasse 25, 6020 Innsbruck-Austria*

ABSTRACT

The most appealing characteristic of PTR-MS is probably the possibility to measure, with high sensitivity, mixtures of volatile compounds in a very fast way. This has been widely exploited in environmental science [1] and medical application [2] and led the food scientists to apply PTR-MS mainly in presence of relatively fast processes: volatile compounds formation during food processing [3,4] or breath analysis and nose–space measurements during and after food consumption [5,6]. Usually in these applications the identification of peaks is not the main issue compared with the possibility to monitor processes in real time and the considered mixtures of volatiles are less complex than those normally found in agroindustrial products.

The absence of separation remains a main limitation of PTR-MS because it does not allow a complete interpretation of the PTR-MS spectra and the methods that have been proposed to overcome this problem [7,8] tend, in our opinion, to reduce its positive characteristics and do not reach the level of identification provided by the very consolidated GC based methods. For this reason static headspace PTR-MS measurements of food related products have not been used widely even if the absence of treatments (no artifacts), the high sensitivity and the possibility to have quantitative indication are often an important advantage as compared to gas chromatography.

Should PTR-MS be used only for the monitoring of relatively fast phenomena or is there any advantage to use it for static assessment of the volatile mixtures characterizing the headspace of food and agroindustrial products? Is a complete identification of peaks necessary or can an unidentified or quasi-unidentified fingerprint be useful, at least in a screening phase?

To address these questions we present here some practical applications concerning agroindustrial applications based on a series of experiments performed in our laboratories. The analytical approach, useful when PTR-MS offers something that more consolidated techniques do not (e.g. PTR-MS is very fast in comparison with gas chromatography), will not be considered here because it has been widely used and described in several publications [3-7, 9] and discussed in contributions to this conference. On the contrary we will show here how, from a methodological point of view, we can take advantage of the PTR-MS spectral fingerprint to characterize the investigated samples. This fingerprint can be used on one hand for classification or identification of samples (e.g. in quality control) and, on the other, to be calibrated with data of other techniques if these are, as it often happens, more expensive or more time consuming than the PTR-MS analysis (e.g. sensory analysis).

PTR-MS spectra being a typical example of high dimensional (many peaks) and inter-correlated (fragmentation, interference) data, their analysis is efficiently carried out by multivariate methods [10]: new variables are built in combining the measured spectral intensity aiming on one hand at a reduction of the variables needed to describe all the "useful" information and, on the other, to build and test models that are able to classify samples (classification) or to predict the value of unknown quantities (calibration). Multivariate analysis of spectroscopic data provides many examples important for our present work and a comprehensive list of references can be found in the related literature [11]. Data mining based on multivariate methods provides a way to manage more efficiently the data analysis problem and extract more information, at least to some extent. Moreover it can indicate the spectral feature that are used by the classification or calibration models helping the analyst in their interpretation.

We will discuss here measurements obtained by a semi-static headspace method that we tested on several kind of food products such as cheese [12,13], fruit juice [14] and fruit [15,16]. Averaged spectra, acquired for a given mass range and resolution time, are used as fingerprint of the samples. Presented data analysis is performed on spectra normalised to unit area.

As an example of classification based on PTR-MS data we show here a few results of a three years project aiming at the characterization of strawberry cultivars. During the first PTR-MS conference [17] we suggested the possibility to recognise the cultivar of a single whole strawberry fruit by non-destructive PTR-MS measurements. Here we show that this indication seems to be confirmed over several years (2002-2004).

Comparing PTR-MS with sensory data [12,18] provides an interesting example of calibration both because of its intrinsic fundamental interest and because the possibility to reduce the use of the time consuming and expensive sensory analysis only to the calibration phase is of applicative and economic importance. Here we present data on hard cheese sensory profile and oil defects [19] concentrating on the description and comparison of different methods. Analogous methods has been applied to the problem of malodor control in waste management plants [20].

In the comparison of molecular data with PTR-MS analysis we found that genetics provides examples of both proposed uses of PTR-MS indicating that this technique could be an useful tool for metabolomic comparative studies: on one side, analytical information can by compared with functional genomic investigation (micro array data of strawberry cultivar [21]), on the other, unidentified fingerprinting is suited to be coupled with molecular markers (apple genotypes [22]).

PTR-MS spectra can be exploited for the analytical information they contain or as a fast and accurate unidentified fingerprint of the investigated samples. In this latter case multivariate analysis allows both the classification of samples and the correlation with other techniques. We believe that coupling PTR-MS spectral fingerprints with suitable data mining techniques provides a new tool that is more powerful than just the sum of the performances of the two separate methods.

Acknowledgments: Work partially supported by the project AGRIIND (PAT, Trento, Italy), RASO (PAT, Trento, Italy) and QUALIFRAPE (MIUR-MURST, Rome, Italy).

References

1. Crutzen, P.J., Williams, J., Pöschl, U., Hoor, P., Fischer, H., Warneke, C., Holzinger, R., Hansel, A., Lindinger, W., Scheeren, B., Lelieveld, J. (2000). High spatial and temporal resolution measurements of primary organics and their oxidation products over the tropical forests of Surinam. Atmospheric Environment 34, 1161-1165.
2. W. Lindinger, J. Taucher, A. Jordan, A. Hansel, and W. Vogel. (1997). Endogenous Production of Methanol after the Consuption of Fruit. Alcoholism: Clinical and Experimental Research vol.21, No. 5 August.
3. Chahan Yeretzian, Alfons Jordan, Raphael Badoud, Werner Lindinger. (2002). From the green bean to the cup of coffee: investigating coffee roasting by on-line monitoring of volatiles. Eur Food Res Technol, 214, 92-104.
4. Philippe Pollien, Christian Lindinger, Chahan Yeretzian, and Imre Blank. (2003). Proton Transfer Reaction Mass Spectrometry, a Tool for On-Line Monitoring of Acrylamide Formation in the Headspace of Maillard Reaction Systems and Processed Food. Analytical Chemistry, Volume 75, Number 20, pp. 5488-5494.
5. Johann Taucher, Armin Hansel, Alfons Jordan, and Werner Lindinger. (1996). Analysis of Compounds in Human Breath after Ingestion of Garlic Using Proton-Transfer-Reaction Mass Spectrometry. J. Agric. Food Chem., 44, 3778-3782.
6. Dagmar Mayr, Tilmann Mär k, Werner Lindinger, Hugues Brevard, Chahan Yeretzian. (2003). Breath-by-breath analysis of banana aroma by proton transfer reaction mass spectrometry. International Journal of Mass Spectrometry 223–224, 743–756.
7. Lindinger W., Hansel A., Jordan A. (1998) On-line monitoring of volatile organic compounds at ppt level by means of Proton-Transfer-Reaction Mass Spectrometry (PTR-MS): Medical application, food control and environmental research. International Journal of Mass Spectrometry and Ion Processes, 173, 191-241.
8. Chahan Yeretzian, Alfons Jordan, Werner Lindinger. (2003). Analysing the headspace of coffee by proton-transfer-reaction mass-spectrometry. International Journal of Mass Spectrometry 223–224 , 115–139.
9. Fall, R.; Karl, T.; Hansel, A.; Jordan, A.; Lindinger, W. (1999). Volatile organic compounds emitted after leaf wounding. On-Line analysis by proton-transfer-reaction mass spectrometry. J. Geophys. Res., 104, 15, 963.
10. Eriksson, L., Johansson, E., Kettaneh-Wold, N., Wold, S. (1999). Introduction to Multi and Megavariate Data Analysis using Projection Methods (PCA & PLS). Umetrics AB, Umeå, Sweden.
11. Kemsley, E.K. Discriminant Analysis and Class Modelling of Spectroscopic Data. John Wiley & Sons Ltd: Chichester,UK, 1998.
12. Gasperi F., Gallerani G., Boschetti A., Biasioli F., Monetti A., Boscaini E., Jordan A., Lindinger W., Iannotta S. (2001) The mozzarella cheese flavour profile: a comparison between judge panel analysis and proton transfer reaction mass spectrometry. J. Sci. Food Agric. 81, 357-363.
13. E. Boscaini, S. Van Ruth, F. Biasioli, F. Gasperi, T. D. Märk, Gas Chromatography-Olfactometry (GC-O) and Proton Transfer Reaction-Mass Spectrometry (PTR-MS) Analysis of the Flavor Profile of Grana Padano, Parmigiano Reggiano, and Grana Trentino Cheeses. (2003). J. Agric. Food Chem. 51, 1782-1790.
14. Biasioli F., Gasperi F., Aprea E., L. Colato, Boscaini E., Märk T. (2003) Fingerprinting mass spectrometry by PTR-MS: heat treatment vs. pressure treatments of red orange juice – a case study. Int. J. Mass Spectrom. 223-224, 343-353.
15. A. Boschetti, F. Biasioli, M. van Opbergen, C. Warneke, A. Jordan, R. Holzinger, P. Prazeller, T. Karl, A. Hansel, W. Lindinger, S. Iannotta. PTR-MS real time monitoring of the emission of volatile organic compounds during postharvest aging of berryfruit. (1999). Postharvest Biology and Technology 17, 143–151.
16. Biasioli F., Gasperi F., Aprea E., Mott D., Boscaini E., Mayr D., Märk T.D. (2003). Coupling Proton Transfer Reaction-Mass Spectrometry with Linear Discriminant Analysis: a Case Study. J. Agric. Food Chem. 51, 7227-7233.
17. Biasioli F., Gasperi F., Aprea E., Mott D., Marini F., Märk T.D. (2003). Discriminant analysis on PTR-MS data for agroindustrial applications. In Contributions to the 1st International Conference on

Proton Transfer Reaction-Mass Spectrometry and Its Applications. Hansel, A., Märk, T.D.. Institut für Ionenphysik, Universität Innsbruck, Austria.

18. Franco Biasioli, Flavia Gasperi, Eugenio Aprea, Isabella Endrizzi,Valeria Framondino, Federico Marini, Daniela Mott, Tilmann D. Märk. (2004). Correlation of PTR-MS spectral fingerprints with sensory characterisation of flavour and odour profile of "Trentingrana" cheese. Submitted to Food Quality and Preferences.

19. Franco Biasioli, Flavia Gasperi, Gino Odorizzi, Eugenio Aprea, Daniela Mott, Federico Marini, Gianmarco Autiero, Giampaolo Rotondo and Tilmann D. Märk. Applicabilità del PTR-MS al controllo degli odori negli impianti per il trattamento dei rifiuti. RS Rifiuti Solidi vol. XVIII n. 4 luglio-agosto 2004.

20. Franco Biasioli, Flavia Gasperi, Gino Odorizzi, Eugenio Aprea, Daniela Mott, Federico Marini, Gianmarco Autiero, Giampaolo Rotondo and Tilmann D. Märk. (2004). PTR-MS monitoring of odour emissions from composting plants. In press. International Journal of Mass Spectrometry.

21. Fabrizio Carbone, Fabienne Mourgues, Franco Biasioli, Flavia Gasperi, Tilmann D.Märk, Gaetano Perrotta, Carlo Rosati. (2004). Comparative analysis of gene expression of strawberry fruit and correlation with profiles of volatile compounds and other quality traits in different genotypes. Poster. This conference

22. Elena Zini, Franco Biasioli, Flavia Gasperi, Daniela Mott, Eugenio Aprea, Tilmann D. Märk, Andrea Patocchi, Cesare Gessler and Matteo Komjanc. (2004). QTL mapping of volatile compounds in ripe apples detected by proton transfer reaction-mass spectrometry. .Poster. This conference

Strawberry flavour analysis by PTR-MS: various low and high time resolution applications

Saskia van Ruth, Katja Buhr, Conor Delahunty

University College Cork, Department of Food and Nutritional Sciences, Western Road, Cork, Ireland, email: s.vanruth@ucc.ie

ABSTRACT

The volatile flavour of strawberries was evaluated using Proton Transfer Reaction Mass Spectrometery (PTR-MS). Low time resolution measurements were carried out on strawberries during maturation. Rapid changes in strawberry flavour composition were measured using real time model mouth analysis and in-nose analysis.

1. Introduction

Strawberry flavour is one of the most complex fruit flavours. Its composition has been extensively studied and more than 300 flavour volatiles have been identified in strawberries [1,2]. Not all of these compounds contribute significantly to the overall flavour, and some authors have used gas chromatography-olfactometry to identify the most important flavour compounds. For instance Larsen et al. [3] reported that ethyl butanoate, 2,5-dimethyl-4-hydroxy-3(2H)-furanone and ethyl hexanoate are character impact flavour compounds. Schieberle [4] selected the first two compounds as key odourants, but also reported (Z)-3-hexenal, methyl butanoate, methyl 2-methylpropanoate and 2,3-butanedione as important flavour compounds. The selections of these important compounds for strawberry flavour were all based on their concentrations in the strawberries. This concentration profile does not necessarily mirror the profile of the compounds released from strawberries and experienced during consumption. The composition of volatiles in the headspace of a food product is generally thought to be closer to what is actually experienced during eating.

PTR-MS is a direct headspace MS technique which allows rapid analyis of volatile flavour compounds, even at pptv concentration levels. It has been successfully applied to measure *in vitro* and *in vivo* flavour release from dried and canned foods during consumption in the form of model mouth analysis and in-nose analysis, respectively [5,6].

In the present study, PTR-MS was explored for measuring slow and rapid changes in the volatile profiles of fresh strawberries. Measurements of relatively slow changes involved fingerprinting MS of strawberry fruits of different maturation degrees. High time resolution analysis concerned real time model mouth analysis and in-nose analysis of strawberries during oral processing.

2. Materials and Methods

Fresh Irish strawberries were purchased in a local supermarket. PTR-MS analyses were carried. Low time resolution analysis involved headspace analysis. For a spectrum of intact strawberries five intact strawberries were placed in a 500 ml glass. For the maturation study, 1 g pieces of strawberries were used. After equilibration for 1 h at room temperature the headspace was sampled at a flow rate of 20 ml/min, 15 ml/min of which was led into the PTR-MS. The headspace was scanned covering the mass range m/z 20-170 using a dwell time of 0.2 s per mass. For model mouth analysis the set up described earlier was used [6], using pieces of 0.5 g fresh strawberry and 0.35 ml of water as saliva substitute. A mastication rate of 52 cycles/min was applied. The headspace of the model mouth was sampled for 2 min at a flow rate of 200 ml/min 15 ml/min of which was led into the PTR-MS. Twenty-four masses were monitored (m/z 21, 32, 33, 37, 45, 59, 61, 75, 83, 85, 87, 89, 99, 103. 111, 115, 117, 129, 131, 137, 145, 153, 155 and 171) resulting in a cycle time of 2.0 s. Their selection was based on preliminary studies and volatile flavour compounds identified in strawberries [1,2]. Four replicates were analysed.

In nose analysis was carried out according to the method described previously [6], using a group of six subjects. Subjects consumed one whole strawberry for each replicate measurement using an imposed oral processing regime. They placed a strawberry in their mouths 20s after the start of the analysis, chewed for 10 s, swallowed at 30 s. During the experiment, the expired breath of the subject was sampled at a flow rate of 100 ml/min, 15 ml/min of which was led into the PTR-MS. The following masses were monitored: m/z 21, 32, 37, 45, 59, 61, 75, 83, 89, 99, 103, 117 and 131. This approach resulted in a cycle time of 1.2 s. Three replicate measurements were carried out for each subject.

3. Results and Discussion

Headspace analysis was carried out on strawberry fruits of various maturation degrees using PTR-MS. Initially a measurement on intact red strawberries was carried out, the resulting spectrum of which is presented in Fig. 1. The two ions with highest intensities observed are in decreasing order m/z 75 and 59. When comparing these data with known fragmentation patterns [7] and previously identified strawberry flavour compounds [1,2], these masses are likely to originate from methyl acetate and acetone/propanal, respectively. For the maturation experiments, fruit pieces were used. The damaging of the strawberry tissue resulted in a change in the spectrum. Mass m/z 45 was the most abundant volatile compound in the headspace of red strawberry pieces. This is likely to be the result of wound response reactions. Green, light green, light red, fully red and dark red strawberries were selected for the maturation study and headspace analyses were carried out. The intensities of the masses m/z 43, 59, 61 and 89 in the headspace of the strawberry pieces increased significantly during maturation. They may reflect some different alcohols (m/z 43), acetone/propanal (m/z 59), acetic acid (m/z 61) and different forms of butyrate or butanoic acid (m/z 89). The intensities of two masses clearly decreased during maturation: mass m/z 33 and m/z 81, which may be related to methanol and furfural, respectively.

Red strawberry pieces were subjected to model mouth analysis. In decreasing order of abundancy, the masses measured were m/z 45, 75, 59, 33/61/99, and 83/89. The change in real time concentrations for three different masses (m/z 33, 61 and 99) are presented in Fig. 2. Three types of release profiles were observed. Type 1 showed after an initial steep increase a relatively constant headspace concentration. This is observed for e.g. mass m/z 33, which may reflect methanol. The Type 2 profiles also reached a plateau concentration, but the initial increase was more gradual. Mass m/z 99 showed this type of behaviour. The more gradual start is likely to relate to formation of the compounds due to wound response reactions. Mass m/z 99 may originate from hexenal, a typical volatile compound formed when fruits and vegetables are damaged. The third profile group demonstrated an initial increase, reached a maximum and showed subsequently a decrease in headspace concentration. Mass m/z 61 is an example of this group, and may reflect acetic acid. This is due to the fact that at a certain stage the rate of flavour transport from the strawberries to the liquid phase and air phase becomes rate limited. Resistance to mass transfer is an important factor determining flavour release. Some compounds may also be transformed into other compounds or may simply be depleted.

In nose analysis was carried out on intact strawberries using a group of 6 volunteers. The change of concentration of mass m/z 75 (probably methyl acetate) in the expired breath of the subjects during eating of a strawberry is presented in Fig. 3. The subjects consumed the strawberries according to a specific protocol. It is remarkable that an increase in concentration of mass m/z 75 was only observed after swallowing. This is, however, a common phenomenon for liquid samples as subjects tend to keep the velum in their throat closed to avoid liquid to go to the nasal cavity [8]. The soft texture of the strawberries and its high water content resulted in fast liquidisation of the food during consumption. This explains the

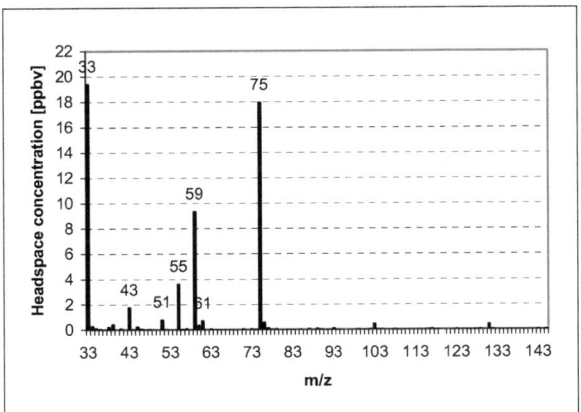

Fig. 1 Spectrum of the volatile compounds of intact red strawberries determined by PTR-MS

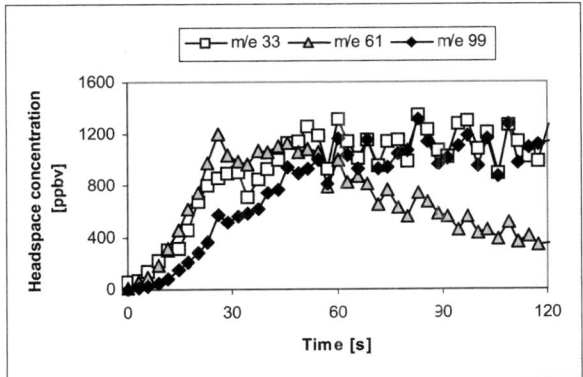

Fig. 2 Real time concentrations of mass m/z 33, 61 and 99 in air during model mouth analysis of red strawberry pieces (four replicates)

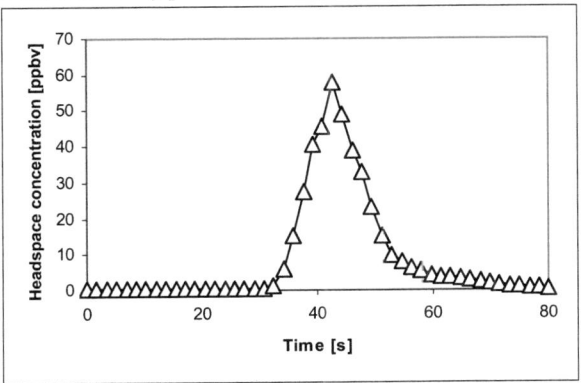

Fig. 3 Real time concentrations of mass m/z 75 in expired breath during in-nose analysis of intact red strawberries (six subjects, three replicates)

oral processing behaviour for liquids. Mass m/z 75 was the most abundant mass found in the expired breath of the subjects. In model mouth analysis this compound was also present at high concentrations, but was second highest after mass m/z 45. Differences between model mouth and in-nose analyses data are due to (1) strawberries pieces in model mouth analysis versus intact strawberries in in-nose analysis were used as samples due to the dimensions of the model mouth; (2) the fact that in in-nose analysis strawberries were swallowed; (3) model mouth analysis mimics release in the mouth. During in-nose analysis the expired breath is sampled at the nostrils. This implies that other physiological factors may influence the concentrations as well which are not taken into account in the model mouth.

4. Conclusions

PTR-MS measured adequately slow and rapid changes in headspace concentrations of volatile compounds emitted from strawberries, i.e. during maturation and during consumption.

5. References

1   R. Tressl, F. Drawert, W. Heiman. Z Naturforschg 21b (1969) 1201.
2   I. Zabatakis, M.A. Holden. J Sci Food Agric 74 (1997) 421.
3   M. Larsen, L. Poll, C.E. Olsen. Z Lebensm Unters Forsch 195 (1992) 536.
4   P. Schieberle, T. Hoffmann. J Agric Food Chem 45 (1997) 227.
5   S.M. van Ruth, K. Buhr, Eur Food Res Technol 216 (2003) 216.
6   S.M. Van Ruth, L. Dings, K. Buhr, M.A. Posthumus. Food Res Int 37 (2004) 785.
7   K. Buhr, S.M. van Ruth. Int J Mass Spectrom 221 (2002) 1.
8   A. Buettner, M. Montserrat. Proceedings of the 7th Weurman Symposium on Flavor Chemistry and Biology (2004), in press.

Generation of Flavour Compounds and Processing Contaminants by Maillard-Type Reactions

Imre Blank and Philippe Pollien

Nestlé Research Center, Vers-chez-les-Blanc, 1000 Lausanne 26, Switzerland
(imre.blank@rdls.nestle.com)

ABSTRACT

The formation of vinylogous compounds and odour-active Strecker aldehydes from amino acids was studied in binary dry mixtures of fructose and valine, asparagine or phenylalanine at 180 °C. The volatile compounds were monitored by PTR-MS. Acrylamide was the major vinylogous compound followed by styrene and traces of 2-methylpropene. On the contrary, methylpropanal was the most abundant Strecker aldehyde followed by phenylacetaldehyde, whereas 3-oxopropanamide could not unequivocally be identified. These data indicate that the decomposition of amino acids to Strecker aldehydes and/or vinylogous compounds *via* Maillard-type reactions cannot be generalized and hardly predicted.

1. Introduction

The recent discovery of relatively high amounts of acrylamide in carbohydrate-rich foods obtained by thermal processing [1, 2] has led to numerous studies indicating Maillard-type reactions as a major reaction pathway, in particular in the presence of asparagine, which directly provides the backbone of acrylamide [3-6]. Similarly, other vinylogous compounds have been identified as reaction products of specific amino acids, such as acrylic acid [7], 3-butenamide [7, 8], and styrene [9] generated from aspartic acid, glutamine, and phenylalanine respectively. The mechanism explaining acrylamide formation from asparagine can basically be applied to other amino acids, as suggested in our recent paper [9]. It is based on a Strecker-type degradation of the Schiff base leading to azomethine ylide intermediates followed by a β-elimination reaction of the decarboxylated Amadori compound to afford the vinylogous compound [9]. Unfortunately, the large majority of the studies dealing with acrylamide and other processing contaminants do not consider flavour or colour formation, despite the fact that they are also formed by Maillard-type reactions implying similar reaction pathways [10]. Therefore, the objective of this study was to simultaneously monitor the formation of some flavour compounds and vinylogous-type processing contaminants using proton transfer reaction mass spectrometry (PTR-MS) [11].

2. Experimental

Materials. L-Valine (Val), L-asparagine (Asn), L-phenylalanine (Phe), D-fructose (Fru), 2-methylpropene, acrylamide, styrene, methylpropanal, and phenylacetaldehyde were from Fluka/Aldrich (Buchs, Switzerland).
Analytical Methods. Samples for on-line measurement of headspace volatiles obtained in pyrolysis experiments were analysed by PTR-MS. The precursors (each 0.35 mol) were ground, mixed, and heated from room temperature to 190 °C at a 5 °C/min heating rate. Acrylamide (*m/z* 72), styrene (*m/z* 105), 2-methylpropene (*m/z* 57), and 3-oxopropanamide

(*m/z* 88), phenylacetaldehyde (*m/z* 121), methylpropanal (*m/z* 73) were monitored in the scan mode (*m/z* 21-200, 0.2 s/mass).

3. Results and Discussion

Vinylogous compounds such as acrylamide have recently been claimed as food processing contaminants and associated with safety risks [1]. This is due to the fact that they can lead to highly reactive epoxides (*i.e.* glycidamide), which may for example react with nucleophiles forming haemoglobin adducts [2]. In analogy to acrylamide **1**, valine and phenylalanine may lead to 2-methylpropene **3** and styrene **5**, respectively (Figure 1). They are formed *via* the Maillard reaction under low-moisture conditions by a Strecker-type degradation of the intermediary Schiff base leading to a decarboxylated Amadori compound that upon β-elimination may release the vinylogous compounds [5-6, 9]. Similar reactions can also lead to Strecker aldehydes, *i.e.* 3-oxopropanamide **2**, methylpropanal **4**, and phenylacetaldehyde **6** (Figure 1), some of which are odour-active contributing to the overall flavour of food products [10, 12].

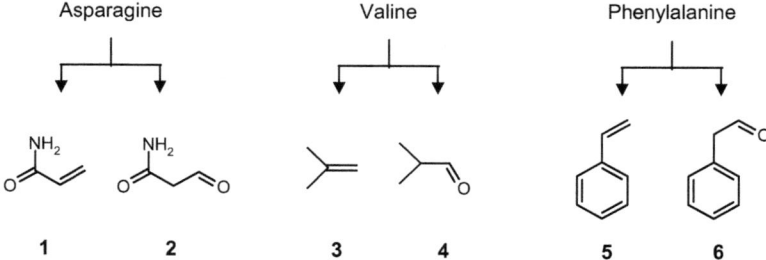

Figure 1. Formation of vinylogous compounds (1, 3, 5) and Strecker aldehydes (2, 4, 6) by Maillard-type reactions (see text for explanation).

PTR-MS was used as a suitable tool to monitor the formation of the compounds **1-6** due to their volatility. As shown in Figure 2A, acrylamide **1** (*m/z* 72) was the major vinylogous compound formed, followed by styrene **5** (*m/z* 105) and traces of 2-methylpropene **3** (*m/z* 57). However, these data are only indicative because the release of compounds into the headspace depends on their volatility. For example, acrylamide **1** is very polar having a high boiling point of 125 °C (25 mm), whereas 2-methylpropene **3** is very apolar and volatile (b.p. -6.9 °C). Styrene **5** is also apolar having a higher boiling point (145 °C) due to π–stacking interactions. Interestingly, **1** shows one major peak corresponding to 150 °C while styrene **5** is basically formed in two portions, *i.e.* at 150 °C and under harsher pyrolytic conditions at 190 °C. The trace at *m/z* 57 is just partially representing **3**, *i.e.* less than 1% corresponds to 2-methylpropene. On the contrary, *m/z* 72 and *m/z* 105 are characteristic for acrylamide **1** and styrene **5**, respectively. Styrene **5** has previously been reported as reaction product of dry and aqueous sugar/Phe systems [13-14].

The formation of the Strecker aldehydes shows a different behaviour (Figure 2B). The malt-like smelling methylpropanal **4** was the most abundant compound followed by the honey-like smelling phenylacetaldehyde **6**, both generated at about 150 °C. Surprisingly, the Strecker aldehyde of asparagine (**2**) was hardly detectable, indicating that this amino acid preferably forms the vinylogous compound. On the contrary, valine favours the generation the Strecker aldehyde. Phenylalanine seems to form both chemical entities under the same conditions. Again, the PTR-MS traces depend on the boiling point of the Strecker aldehydes, *i.e.* 63 °C and 195 °C for **4** and **6**, respectively.

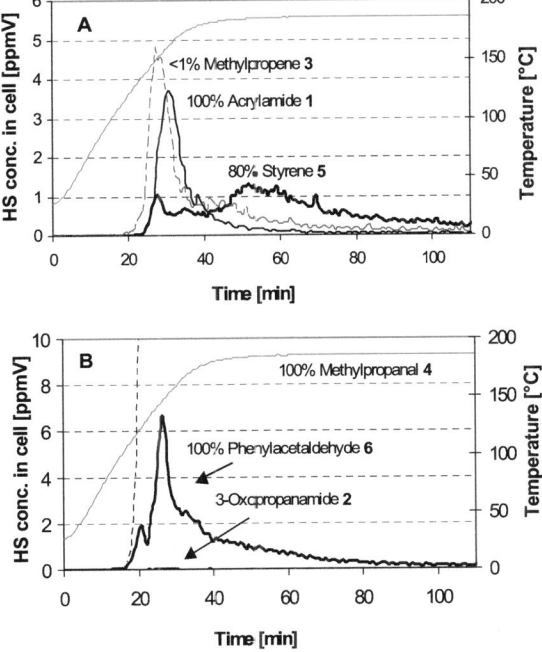

Figure 2. Formation of (A) vinylogous compounds: acrylamide **1**, 2-methylpropene **3**, styrene **5** and (B) Strecker aldehydes: 3-oxopropanamide **2**, methylpropanal **4**, phenylacetaldehyde **6**.

All attempts to identify substantial amounts of the Strecker aldehyde 3-oxopropanamide failed so far. As shown in Figure 3, even on-line measuring tools based on PTR-MS indicated only traces of a compound with m/z 88 ($C_3H_5NO_2$, protonated). Possible fragments pointing to the Strecker aldehyde could neither be found, *i.e.* $[M+1-NH_3]^+$ (m/z 71) and $[M+1-H_2O]^+$ (m/z 70). However, the signal representing acrylamide (m/z 72) formed under the same reaction conditions could easily be monitored. These data suggest that acrylamide is preferably generated compared to the Strecker aldehyde. Interestingly, no published data could be found on 3-oxopropanamide as Strecker aldehyde of asparagine using SciFinder®. In general, it has only been mentioned in a few papers, mainly dealing with computational chemistry related to intramolecular N-H···O resonance-assisted hydrogen bonding in β-enaminones [15, 16]. In fact, 3-oxopropanamide may preferably occur as 3-amino-3-hydroxy-2-propenal stabilized by intramolecular N-H···O bonding.

4. Conclusions

PTR-MS turned out to be a suitable analytical tool for on-line monitoring of flavour compounds and processing contaminants generated upon heat treatment. Strecker aldehydes and vinylogous compounds show different formation patterns, suggesting that these molecules are generated by various reaction pathways. Therefore, it should be possible to favour the formation of flavour-active components while controlling the amounts of vinylogous compounds. However, more work is required to develop food products with desirable flavour notes and reduced amounts of processing contaminants.

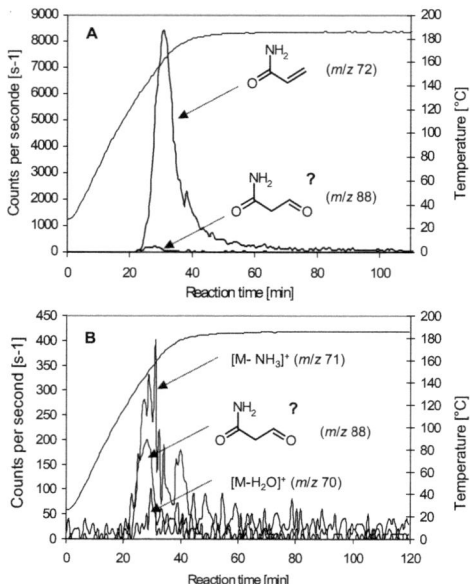

Figure 3. (A) Formation of acrylamide (*m/z* 72) and putative 3-oxopropanamide (*m/z* 88) from binary mixtures of fructose and asparagine monohydrate monitored by PTR-MS. (B) Traces of potential fragments of putative 3-oxopropanamide

5. References

[1] E. Tareke, P. Rydberg, P. Karlsson, S. Eriksson, and M. Törnqvist, *J. Agric. Food Chem.* 2002, **50**, 4998.

[2] M. Friedman, *J. Agric. Food Chem.* 2003, **51**, 4504.

[3] D. S. Mottram, B. L. Wedzicha, and A. T. Dodson, *Nature* 2002, **419**, 448.

[4] R. H. Stadler, I. Blank, N. Varga, F. Robert, J. Hau, Ph. A. Guy, M.-C. Robert, and S. Riediker, *Nature* 2002, **419**, 449.

[5] V. A. Yaylayan, A. Wnorowski, and C. Perez Locas, *J. Agric. Food Chem.* 2003, **51**, 1753.

[6] D. V. Zyzak, R. A. Sanders, M. Stojanovic, D. H Tallmadge, B. L. Eberhart, D. K. Ewald, D. C. Gruber, T. R. Morsch, M. A. Strothers, G. P. Rizzi, and M. D. Villagran, *J. Agric. Food Chem.* 2003, **51**, 4782.

[7] R. H. Stadler, L. Verzegnassi, N. Varga, M. Grigorov, A. Studer, S. Riediker, and B. Schilter, *Chem. Res. Tox.* 2003, **16**, 1242.

[8] R. Weißhaar and B. Gutsche, Deuts*che Lebensm. Rundsch.* 2002, **98**, 397.

[9] R. H. Stadler, F. Robert, S. Riediker, N. Varga, T. Davidek, S. Devaud, T. Goldmann, and I. Blank, *J. Agric. Food Chem.* 2004, **52**, 5550.

[10] F. Ledl and E. Schleicher, *Angew. Chem. Int. Ed. Engl.* 1990, **29**, 565.

[11] P. Pollien, C. Lindinger, C. Yeretzian, and I. Blank, *Anal. Chem.* 2003, **75**, 5488.

[12] H. Weenen and J. G. M van der Ven, In: G. R. Takeoka, M. Güntert, K.-H. Engel (Eds.), Aroma Active Compounds in Foods, American Chemical Society, Washington, 2001, p. 183.

[13] A. Keyhani and V. A. Yaylayan, *J. Agric. Food Chem.* 1996, **44**, 223.

[14] G. Westphal and E. Cieślik, *Nahrung* 1982, **26**, 765.

[15] P. Gilli, V. Bertolasi, V. Ferretti, and G. Gilli, *J. Am. Chem. Soc.* 2000, **122**, 10405.

[16] M. A. Rios and J. Rodriguez, *Theochem.* 1991, **74**, 149.

Direct inter-comparison of datasets obtained by different PTR-MS: A novel approach to optimize and adapt the fragmentation pattern using a standardization procedure

Christian Lindinger[1,2], Philippe Pollien[1], Santo Ali[1], Imre Blank[1] and Tilmann Märk[2]

[1] *Nestlé Research Center, Vers-chez-les-Blanc, 1000 Lausanne 26, Switzerland*

[2] *Institut für Ionenphysik, Leopold Franzens Universität Innsbruck, 6020 Innsbruck, Austria*

Monitoring volatile organic compounds released by different food matrices during manufacturing and packaging processes is one of the strengths of PTR-MS because of its on-line capability. Once the key compounds are identified their relative abundance can be continuously analysed for quality control and safety purposes. An application in the production of coffee is the online product characterization during the roasting process and analytic aroma profiling of coffee headspace. Here absolute quantification is often not needed as the relative abundance of different flavour compounds and their precursors is essential. As soon as absolute quantification is necessary to obtain quality certification, PTR-MS has a great potential but still there are some important weaknesses. For absolute quantification information about fragmentation patterns, reaction rate constants, reaction time, drift tube pressure and temperature, humidity of the buffer gas and transmission characteristics of the quadrupole mass spectrometer is needed. Especially when data sets obtained with different PTR-MS instruments are compared, it is necessary to standardize all instrumental parameters affecting product ion abundance.

Here we propose a new method to:

- measure the quadrupole transmission characteristics in a fully automated manner in the range from m/z 20 to m/z 200
- automatically set the instrumental parameters to allow for direct comparison of data sets obtained with different PTR-MS devices

The developed approach solves the problem of inter-instrumental variability and will allow for the development of a large spectra evaluation database.

Biochemistry and physical chemistry of flavour in plant tissues and fabricated food systems: Examples and approaches using PTR-MS

Patrick Dunphy, Ian Butler and Ingmar Qvist

Danisco UK Ltd., Denington Road, Wellingborough, Northants., NN8 2QJ, UK.
(patrick.dunphy@danisco.com)

ABSTRACT

The capability to monitor quantitatively and in real time flavour dynamics using PTR-MS has opened up a number of new avenues for exploration. These include aspects of plant flavour biochemistry, physical chemistry of flavour release from structured fabricated foods, insights into the release of aroma compounds during the eating and drinking process and linking of in-nose aroma concentration, via retro-nasal flavour thresholds, to flavour perception. This paper will demonstrate this potential for PTR-MS by examples drawn from the above areas and in addition illustrate how this technique in combination with other flavour technologies can lead to new opportunities in flavour development.

1. Introduction

Chemical ionization mass spectrometric methods such as Proton Transfer Reaction Mass Spectrometry (PTR-MS) and Atmospheric Pressure Chemical Ionization Mass Spectrometry (APCI) were recently applied to the field of foods and flavour research (1,2,3). The SDT-PTR-MS in particular provides the opportunity for gas phase quantification and greater sensitivity, in the range ppbv-ppmv, and therefore was the instrument of choice in these studies.

This capability to monitor in real time gas phase concentrations of aroma compounds has opened up opportunities for real time monitoring of flavour formation and release in systems as diverse as wounded plant tissues and in nose aroma release from fabricated and natural food systems during consumption. This paper exemplifies some of the wide range of applications possible with this technology.

2. Applications

Plant flavour biochemistry
Many of the compounds that contribute to the characteristic aroma of fruit are known to be formed *de novo* in response to induced stress including excessive heating, cooling, dehydration, predator, insect, microbial attack and mechanical damage (4). In response to such incursions plants initiate a chain of reactions, among which is the oxylipin cascade, this resulting in the formation of C_6 aldehydes and related transformation products by enzymatic oxidation and C-C scission of the intermediates hydroperoxides of the precursors linoleic and linolenic acids. The most aroma potent and first formed volatile compound produced is the unstable (z)-3-hexenal. The transient nature of this cascade contrasts with the pattern of release observed for acetaldehyde under the same conditions of tissue damage. This C_2 compound exhibits a pattern of release that slowly builds up and is sustained over a time period when the C_6 aldehydes have maximized then declined. This phenomenon has been observed in damaged apple and strawberry fruit and appears to be part of the wound response reaction of the tissue to mechanical damage. Potential routes to acetaldehyde include:
  ➢ Strecker/enzyme type degradation of the amino acid alanine
  ➢ Hydrolysis of acetaldehyde derivatised as the acetal
  ➢ Glycolytic conversion of glucose/fructose to pyruvate and fermentative production of acetaldehyde and ethanol
Experiments are underway to define the pathway(s) involved

Flavour release and its control in fabricated foods

Controlling flavour release from food systems can in principle be achieved by the two main routes of flavour manipulation and/or control of the assembly of the food microstructure and its dis-assembly in the oral cavity. The former approach has limitations since any attempt to change the physico-chemical properties of molecules and thus their release characteristics is usually associated with a change in organoleptic character (Compare the aroma character of the homologous series of straight chain 2-ketones as the carbon number is increased). The second approach on the other hand offers more possibilities in terms of ingredient selection and product fabrication and the dis-assembly of the latter during consumption. Lipid based systems such as oil in water emulsions exemplify the route for influencing flavour release under in- mouth conditions. The classic work of Malone *et al* (5) on the control of flavour release in low fat emulsions demonstrated, both analytically, in nose and sensorially that by understanding the food structural factors and the interaction of the food matrix with the oral cavity that tailoring of the release of lipophilic molecules can be achieved. Work in our own laboratory has demonstrated the effect of addition of alcohol and fat and increasing structural complexity on the release of flavour from beverage matrices and other food systems.

In nose release of aroma compounds during food consumption

Flavour related physiology and anatomy of the oral cavity

The anatomy of the oral cavity and the naso-and oro-pharynx relevant to the eating/swallowing process and the routes of ortho-nasal and retro-nasal flavour delivery are now fairly well understood. Buccal or pre-retro-nasal perception is associated with the temporary and partial opening and closing of the soft palate (velopharangeal closure) which enabled vapours to pass from the mouth in the direction of the nasal cavity, a process usually associated with the consumption of solid food. Full retro-nasal perception is associated with the post-swallowing process itself when the velopharangeal closure fully opens following previous closing and opening of the epiglottis (6,7,8). Using magnetic resonance imaging (MRI) it was possible to confirm the role of the oral cavity during food mastication and swallowing (7). Further studies additionally demonstrated that the breath volatiles profile was also influenced by other mouth movements such as chewing (9). Simultaneous real-time monitoring of mastication, swallowing, nasal airflow and aroma release confirmed that chewing affected nasal airflow with the flow fluctuations following the mastication pattern. A mean volume of 26 ml air was transferred from the mouth into the pharynx with each chew. During swallowing there was no nasal airflow but post -swallowing aroma release was observed (10). It is important however in the overall flavour context however to note that retro-nasal perception receives principal contributions from taste as well as aroma (11).

Volatiles release during eating and swallowing of liquid and solid foods

Using the dynamic capability of PTR-MS in the in-nose mode it was possible to demonstrate the release during consumption of a range of differently compliant foods from beverages through to firm fabricated, confectionery products.

Ethyl butanoate was was used as a marker compound to show the quantitative aroma patterns associated with eating these different materials.

Consumption of 4 ml of a strawberry flavoured beverage containing ethyl butanoate was monitored in the normal drinking mode when the sample is taken into the mouth, then immediately swallowed. Breathing pattern was recorded using isoprene as a breath marker. Only in the post-swallowing mode was there any aroma transfer from the food/oral cavity to the olfactory region. Additional, aftertaste lesser transfer of aroma occurs after a subsequent swallowing, this removing residual ethyl butanoate/saliva mix.

Consumption of a solid food such as a fruit chew produces a much more complex aroma release profile in the timescale, of ca. 50 seconds, of consumption.In addition to the aroma pulses due to the swallowing/post-swallowing cycles there were additional inter-swallowing pulses brought about by the mouth chewing action. This meant that whilst there was "chewable" material in the mouth transfer of aroma-laden air to the nasal cavity occurred.These observations could have consequences for the flavour adaptation process.

The above complex fruit chew product represents a fabricated structure of some resistance to chewing with a prolonged aroma delivery profile. Thus the nature of the food product itself can have a significant influence on flavour release.

Based on these and other studies it was possible to draw up a simple model for the swallowing/aroma release process. For normal consumption of liquids volatile transfer to the olfactory epithelium occurs only at the post-swallowing phase.Oral to nasal cavity gas transfer occurs when the soft palate subsequently closes off the rear of the oral cavity, this coinciding with the post-swallowing phase of consumption of food, the opening of the pharyngeal/nasal cavity interface and culminating in the transfer of flavour laden gas to the olfactory region..The consumption of a solid food and the related oral to nasal gas transfer is more complex and has two distinct phases. The first, so-called buccal transfer occurs with the intermittent open/closing of the soft palate allowing, in synchrony with chewing action, transfer of gas from the mouth to the nasal cavity. The second phase is initiated by the full opening of the soft palate in conjunction with food swallowing with the temporary restriction of the pharyngeal/nasal cavity interface and closure of the epiglottis and is a repeat of the process that occurs during swallowing of a liquid.

In-nose aroma concentrations and retro-nasal flavour threshold values

The ability, by PTR-MS, to measure in real time the in-nose concentration of aroma compounds permits the conversion of this data into relative perceptual intensity values based on a knowledge of retro-nasal thresholds. Consideration of the wide range of threshold values for even similar structures makes it essential when considering the sensory impact of particular aroma compounds to relate measured concentrations to thresholds. This can be demonstrated by consideration of the retro-nasal flavour thresholds for the family of C_6 aldehydes (12). (See Table 1)

compound	Odour Quality	Thr[1]	Rel[2]
(Z)-3-hexenal	Leaf like	0.03	6666
(E)-2-hexenal	Apple like	27	7
(Z)-3-hexenol	Leaf like	30	7
hexanal	Tallowy, leaf like	16	13
hexanol	Green, flowery	200	1

Table1. Flavour threshold and relative flavour intensity values for a group of related C_6 aldehydes.Key: Thr[1] is retro-nasal threshold value in µg / L water. Rel[2] is the retro-nasal threshold value relative to hexanol

Retronasal threshold values were employed since these are measured when the sample under assessment is in the mouth i.e.under the most realistic eating conditions.

From this data it is clear that for equi-weight amounts of (Z)-3-hexenal, (E)-2-hexenal and hexanal the compounds are 6666, 7 and 13 times as odour potent as hexanol.

Using the in nose concentrations and the known threshold values the so called Nasal Intensity Flavour values (NIF's) for individual compounds can be determined as follows :

$$NIF = [concentration]_{in\,nose} / \text{retronasal threshold}$$

This approach bears similarity to currently derived odour activity valued (AOV's). In this way aroma compounds measured in-nose from a particular food system can be ordered in terms of their a heirarchical impact contribution.It should be bourne in mind that retro-nasal values represent an average measured on individuals or small groups and that the methods employed may vary by experimenter.In spite of such limitations a display of information as NIF values is of much greater relevence to perception of aroma than gas phase concentrations alone. Employment of this approach to individuals can realise specific rather than averaged NIF data.

The Commonsense[TM] Flavour Approach

To date flavour development has been driven by the classical flavour science disciplines of isolation, identification and synthesis realising many key aroma compounds such as furaneol and the pyrazines. The procedures employed above generally are followed without detailed consideration of the biochemical and chemical time-frame, pre- or post-damage of the tissue, and thus more likely represented a summation of the individual time-frames. Consideration of the complex biochemistry and physical chemistry of aroma compounds in plant tissue such as fruit during comminution by processing in the factory, consumption in the mouth and *in vitro* confirmed that the processes involved are both dynamic and rapid.It is also critical to note that key components of these rapid biochemical changes are perceived in the timescale of consuming fruit.Thus flavours should reflect where possible dynamics segments of the release process rather than a single, averaged, overall picture. Any future flavour developments should take these factors into account and in fact the approach taken to flavouring, as outlined below, focuses on the flavour composition of fruit flavour pre- and post-damaged covering the time-frame of the dynamics of both *in vitro* and in-mouth fruit consumption. Flavours development therefore was directed to produce a range of **Commonsense**[TM] **Flavours** based on the key contributory disciplines of classical flavour science, flavour dynamics of eating, flavour physical chemistry, flavour biochemistry and the cutting edge technology of PTR-MS. The integration of these technologies, their interactions and dynamics in particular with in-nose aroma release assessment permitted the development of a novel approach to flavour formulation which will be exemplified by reference to strawberry flavour.

3. References

1. Lindinger, W., Hansel, A. and Jordan, A., Int. J. Mass Spectrom. Ion Proc., 173, **1998**, 191-241.
2. Boschetti, A., Biasioli, F., van Opbergen, M., Warnake, C., Jordan, A., Holzinger, R., Prazeller, P., Karl, T., Hansel, A., Lindinger, W. and Iannotta, S., Postharvest Biol. Technol., 17, **1999**, 143-151.
3. Taylor, A. J. and Linforth, R. S. T., Int. J. Mass Spectrom., 223-224, **2003**, 179-191.
4. Blee, E., Prog. Lipid Res., 37, **1998**, 33-72.
5. Malone, M.E., Appelqvist, I.A.M., Goff, T.C., Homan, J.E., and Wilkins, J.P.G. In "Flavour Release". Ed. by Roberts, D.D. and Taylor, A.J. ACS Symposium Series #763, **2000**, Chapter 18, pp. 212-227.
6. Buettner, A., Beer, A., Hannig, C. and Settles, M., Chem. Senses, 26, **2001**, 1211-1219.
7. Buettner, A., Beer, A., Hannig, C., Settles, M. and Schieberle, P., Food Qual. Pref., 13, **2002**, 497-504.
8. Petka, J., Cacho, J. and Ferreira, V., Abstract from Actualites Œnologiques 2003, VIIème Symposium International d'Œnologie, Bordeaux, 19-21 Juin, **2003.**
9. Linforth, R., Hodgson, M. and Taylor, A., "Studies of retronasal flavour delivery". 10th Weurman Flavour Research Symposium, 24-28 June, **2002**, Beaune, France. Intercept Ltd., pp. 143-147.
10. Hodgson, M., Linforth, R.S.T. and Taylor, A.J., J. Agric. Food Chem., 51, **2003**, 5052-5057.
11. Delwiche, J., Food Qual. Pref., 15, **2004**, 137-146.
12. Rychlik, M., Schieberle, P. and Grosch, W. "Compilation of odor thresholds, odor qualities and retention indices of key food odorants." Deutsche Forschung. fur Lebensmittelchemie und Institut für Lebensmittelchemie der Technischen Universitat Munchen, **1998**, 1-63.

2.3 Medical Applications

GC-PTR-MS exploration of UV-induced skin damage

Marco M.L. Steeghs, Bas W.M. Moeskops, Simona M. Cristescu, Karen van Swam and Frans J. M. Harren

Molecular and Laser Physics, Institute for Molecules and Materials, Radboud University Nijmegen, Toernooiveld ,1 6525 ED Nijmegen, The Netherlands.
m.steeghs@science.ru.nl

Trace gas analysis of the breath composition gives information about various processes occurring inside the human body. One such process is lipid peroxidation, in which free radicals induce oxidative degradation of the polyunsaturated fatty acids, causing cell damage and cell death. As markers of free radical-mediated damage in the human body, the measurement of the exhaled hydrocarbons ethylene (C_2H_4), ethane (C_2H_6) and pentane (C_5H_{12}) represent a good and non-invasive method to monitor lipid peroxidation.

Using laser-based photoacoustic trace gas detection, the effect of UV-radiation on the human skin has been monitored by on-line monitoring ethylene emissions, directly from the skin. To gain more insight in the effects of UV-induced lipid peroxidation and the mechanisms behind the damage, PTR-MS analysis of volatile products coming directly from the skin has been performed. This revealed several products of LPO showing similar trends as found for ethylene. Identification of these compounds can be done by coupling a Gas Chromatograph (GC) to the PTR-MS system. Here, the first results of skin-cell measurements using (GC)-PTR-MS on UV-induced lipid peroxidation on the human skin are presented.

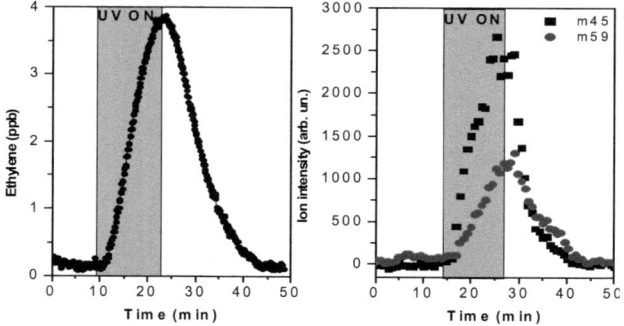

Fig1. On-line measurement of UV-induced lipid peroxidation products using Laser-based photoacoustic trace gas detection (left, ethylene) and PTR-MS (right)

PTR-MS has been used to monitor volatile products of UV-induced damage to the skin. For this, direct skin measurements have been performed using a so called skin cell. Besides ethylene, which is a known biomarker for LPO (1), several products are expected to be produced. In fig.1, right panel, a typical measurement of two of the most abundant masses can be seen. These masses are expected to correspond to acetaldehyde and propanal (propionaldehyde). To validate these and other identifications, GC-PTR-MS measurements will be performed.

(1) H.W.A. Berkelmans, B.W.M. Moeskops, J. Bominaar, P.T.J. Scheepers and F.J.M. Harren Toxicology and Applied Pharmacology 190 (2003) 206-213

Analysis of volatile organic compounds in the headspace of fluid obtained from the gut during colonoscopy by Proton-Transfer-Reaction-Mass Spectrometry – A novel diagnostic approach

M. Lechner[1], H. P. Colvin[1], C. Ginzel[1], P. Lirk[1], H. Tilg[2] , J. Rieder[1]

[1]Department of Anaesthesiology and Critical Care Medicine, University Hospital Innsbruck, Innsbruck, Austria

[2]Department of Medicine, Division of Gastroenterology and Hepatology, University Hospital Innsbruck, Innsbruck, Austria

BACKGROUND:
The diagnosis of many gastro-intestinal diseases is difficult and can often be confirmed only by using invasive diagnostic means. In contrast, the analysis of volatile organic compounds in the headspace of fluid which is obtained from the gut during colonoscopy and in exhaled air may be a novel approach for the diagnosis of these diseases.

MATERIAL AND METHODS:
The volatile compounds were analyzed using Proton-Transfer-Reaction-Mass-Spectrometry (PTR-MS) which allows sensitive and rapid measurement of these substances. Fluid samples obtained from the gut during colonoscopy were collected from 76 and breath samples from 72 subjects. First of all, a profile of volatile organic compounds (VOC) in the headspace of the fluid and in the exhaled air of healthy controls was created. Afterwards this VOC-pattern was compared with that of patients suffering from inflammatory bowel diseases (IBD; Crohn's disease and ulcerative colitis; n=10) and irritable bowel syndrome (IBS; n=7).

RESULTS:
Significant differences in the VOC profile could be observed both in the headspace of the fluid and in the exhaled air comparing patients with healthy controls. However, further studies are necessary to find out whether these differences are specific enough to possibly establish this novel method in clinical routine.

CONCLUSIONS:
VOC-patterns in the headspace of fluid obtained from the gut during colonoscopy could become a novel diagnostic tool in the differential diagnosis of gastro-intestinal diseases.

Breath gas as biochemical probe in sleeping individuals

Anton Amann[1], Stefan Telser[2], Leonhard Hofer[2], Alex Schmid[1,2] and Hartmann Hinterhuber[2]

[1]*Department of Anesthesia and General Intensive Care, Medical University of Innsbruck, A-6020 Innsbruck, Austria, e-mail: anton.amann@uibk.ac.at*

[2]*Department of Psychiatry, Sleep Laboratory, Medical University of Innsbruck, A-6020 Innsbruck, Austria*

ABSTRACT
Breath gas analysis is an emerging scientific field, which has great potential for clinical diagnosis and therapeutic monitoring. Since it is non-invasive with the possibility of online measurements, (restricted) information about the biochemical processes in the sleeping body is available. Extensive quality control is still missing: Room air contaminants, influences of the hemodynamics and lung mechanics, or measurement artefacts can lead to results which are easily misinterpreted. In addition, bacterial metabolization of food in the gut with production of volatiles observed in breath has to be taken into account.
Preliminary results on the volume fraction time series of two compounds, methanol and isoprene, are presented. Isoprene volume fraction has a tendency to increase overnight, whereas methanol shows a decrease, which seems to depend on the chosen protocoll (normal night versus night after sleep deprivation).

1. Introduction
The physical, biochemical and molecular biological methods of medical diagnostics have been developed very rapidly in the recent decades. The main objective has been laid on blood and urine diagnostics. The diagnostics based on human exhaled breath, on the other hand, is much less developed and not widely utilized in clinical practice. Nevertheless, breath gas analysis is an emerging scientific field [1-21], which has great potential for clinical diagnosis and therapeutic monitoring.

Volatile organic compounds in breath are produced by metabolic processes in the body, by bacteria in the gut [22], or both. Isoprene, for example, is a component which is (not necessarily exclusively) produced from dimethyl allyl pyrophosphate, an intermediate compound in the cholesterine synthesis pathway [23]. For a few substances, the biochemical origin is known. For most of the approximately 3000 substances detected in different persons' breath [24], biochemical background information is not available. Some of these substances result from inhalation of contaminants in room air, jet fuel [25] or gasoline.

There is evidence of a biochemical feedback between the sleep-controlling brain and metabolic activities outside the brain [26-31]. Perturbed feedback may be a reason for sleep disorders. Online mass spectroscopic analysis of volatile organic compounds (VOCs) in exhaled air is expected to provide insight into the metabolic processes and hence to complement the information about brain activity obtained by polysomnography (PSG) [4].

2. Methods
Our experimental setup consisting of a proton-transfer-reaction mass spectrometer (PTR-MS, Ionicon FDT-s) combined with PSG (Nihon-Kohden EEG 4317F) was designed for

simultaneous online monitoring of the sleepers' exhaled breath and electrophysiological sleep variables. Exhaled air was continuously sampled through a catheter in the nasal cavity and conducted to the PTR-MS through perfluoroalkoxy copolymer tubing heated to 43°C. The reported masses are those of the protonated species (molecular mass + 1 u) according to the ionization process used in PTR-MS. The concentration time series are partly smoothed using moving average techniques. Sleep stages were determined according to the rules of Rechtschaffen and Kales. Breathing and pulse frequency (both smoothed) were derived from the recorded thorax excursion and ECG data. All participating test persons, 10 healthy men aged 20 to 28, spent 3 nights in the sleep laboratory. The first night was spent there for the purpose of getting familiar with the experimental setup (adaptation night). Subsequently the test persons were deprived from sleep for circa 40 h, i.e., they had to skip a night of sleep, before returning to the laboratory for sleeping (night after sleep deprivation = recovery night). Thereafter, they underwent their daily routine. In the subsequent night, they returned again to the laboratory and slept there for the third and last time (normal night).

3. Results

Isoprene (mass 69u) volume fraction is influenced by pulse frequency, breathing frequency and breathing volume (not shown). It tends to increase during night (fig 1). There is no obvious difference between normal nights and recovery nights. Methanol (33u) concentration decreases during normal nights, whereas it stays rather constant during recovery nights. The ratio of acetaldehyde (45u) to ethanol (47u) concentrations increases during night (not shown).

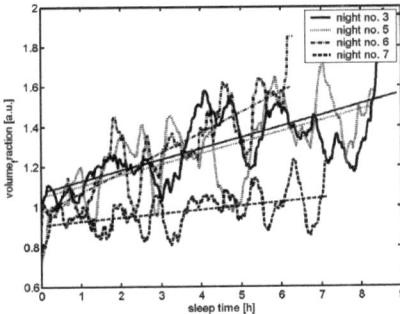

Fig.1: Normalized volume fraction time series for mass 69 (isoprene) for different test persons in normal nights. Normalization to 1 refers to 30 min sleep time. Four different representative nights have been chosen for illustration.

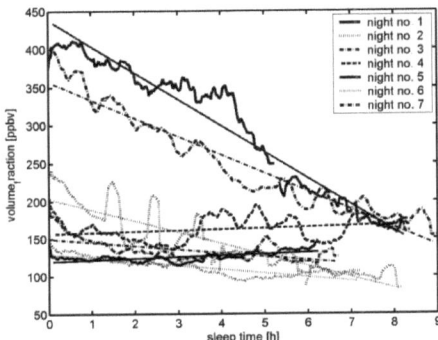

Fig.2: Volume fraction time series for mass 33 (methanol) for different test persons in "normal" nights. If the volume fraction is high, it shows a tendency to decrease; if the volume fraction is already low, it shows a tendendy to stay constant.

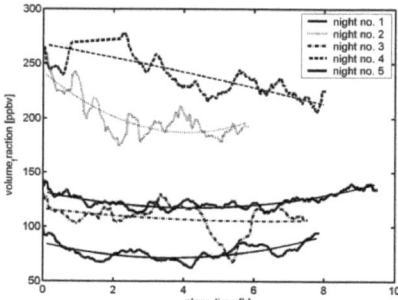

Fig.3: Volume fraction time series for mass 33 (methanol) for different test persons in nights after sleep deprivation. Volume fractions decrease slower than in normal nights or remain approximately constant.

4. Discussion

The evolving concentrations of VOCs such as acetone, methanol, ethanol, acetaldehyde or isoprene reflect biochemical processes. Sometimes it is more convenient to consider derived quantities: for example, the time evolution of the ratio of acetaldehyde to ethanol concentrations reflects the alcohol dehydrogenase activity during the night. The concentration course of isoprene is special, as its concentration varies with pulse frequency and depends on breathing.

The reported results may be spoiled by wrong concentration values, and are thus preliminary observations. In fact, concentration measuring may be influenced by condensed water droplets selectively dessolving hydrophilic compounds; or the tubing between probands and PTR-MS may selectively absorb and desorb certain substances.

Breath gas analysis in sleeping individuals is non-invasive and enables us to monitor metabolic processes on-line, thus complementing the electrophysiological information obtained by polysomnography. Therefore, we believe that it has great potential as a source of additional information to standard polysomnography.

5. References

1. Phillips, M., et al., *Volatile markers of breast cancer in the breath.* Breast J, 2003. **9**(3): p. 184-91.

2. Phillips, M., et al., *Detection of lung cancer with volatile markers in the breath.* Chest, 2003. **123**(6): p. 2115-23.

3. Phillips, M., et al., *Effect of oxygen on breath markers of oxidative stress.* Eur Respir J, 2003. **21**(1): p. 48-51.

4. Amann, A., et al., *Applications of breath gas analysis in medicine.* Int J Mass Spectrometry, 2003: p. to appear.

5. Phillips, M., *Breath tests in medicine.* Sci Am, 1992. **267**(1): p. 74-9.

6. Cope, K.A., et al., *Effects of Ventilation on the Collection of Exhaled Breath in Humans.* J Appl Physiol, 2003.

7. Risby, T.H., et al., *Breath ethane as a marker of reactive oxygen species during manipulation of diet and oxygen tension in rats.* J Appl Physiol, 1999. **86**(2): p. 617-22.

8. Risby, T.H. and S.S. Sehnert, *Clinical application of breath biomarkers of oxidative stress status.* Free Radic Biol Med, 1999. **27**(11-12): p. 1182-92.

9. Risby, T.H., et al., *Breath ethane as a marker of reactive oxygen species during manipulation of diet and oxygen tension in rats.* J Appl Physiol, 1999. **86**(2): p. 617-22.

10. Risby, T.H., *Volatile Organic Compounds as Markers in Normal and Diseased States*, in *Disease Markers in Exhaled Breath*, N. Marczin and M.H. Yacoub, Editors. 2002, IOS Press.

11. Andreoni, K.A., et al., *Ethane: a marker of lipid peroxidation during cardiopulmonary bypass in humans.* Free Radic Biol Med, 1999. **26**(3-4): p. 439-45.

12. Scholpp, J., et al., *Breath markers and soluble lipid peroxidation markers in critically ill patients.* Clin Chem Lab Med, 2002. **40**(6): p. 587-94.

13. Schubert, J.K. and G.F. Noldge-Schomburg, *[Can the limits of intensive care management be defined?].* Zentralbl Chir, 2001. **126**(9): p. 717-21.

14. Miekisch, W., et al., *Analysis of volatile disease markers in blood.* Clin Chem, 2001. **47**(6): p. 1053-60.

15. Schubert, J.K., et al., *CO(2)-controlled sampling of alveolar gas in mechanically ventilated patients.* J Appl Physiol, 2001. **90**(2): p. 486-92.

16. Schubert, J.K. and K. Geiger, *[Importance and perspectives of breath analysis].* Anasthesiol Intensivmed Notfallmed Schmerzther, 1999. **34**(7): p. 391-5.

17. Schubert, J.K., et al., *In vivo evaluation of a new method for chemical analysis of volatile components in the respiratory gas of mechanically ventilated patients.* Technol Health Care, 1999. **7**(1): p. 29-37.

18. Diskin, A.M., P. Spanel, and D. Smith, *Time variation of ammonia, acetone, isoprene and ethanol in breath: a quantitative SIFT-MS study over 30 days.* Physiol Meas, 2003. **24**(1): p. 107-19.

19. Diskin, A.M., P. Spanel, and D. Smith, *Increase of acetone and ammonia in urine headspace and breath during ovulation quantified using selected ion flow tube mass spectrometry.* Physiol Meas, 2003. **24**(1): p. 191-9.

20. Smith, D., T. Wang, and P. Spanel, *On-line, simultaneous quantification of ethanol, some metabolites and water vapour in breath following the ingestion of alcohol.* Physiol Meas, 2002. **23**(3): p. 477-89.

21. Smith, D., et al., *Comparative measurements of total body water in healthy volunteers by online breath deuterium measurement and other near-subject methods.* Am J Clin Nutr, 2002. **76**(6): p. 1295-301.

22. Romagnuolo, J., D. Schiller, and R.J. Bailey, *Using breath tests wisely in a gastroenterology practice: an evidence-based review of indications and pitfalls in interpretation.* Am J Gastroenterol, 2002. **97**(5): p. 1113-26.

23. Sharkey, T.D., *Isoprene synthesis by plants and animals.* Endeavour, 1996. **20**(2): p. 74-8.

24. Phillips, M., et al., *Variation in volatile organic compounds in the breath of normal humans.* J Chromatogr B Biomed Sci Appl, 1999. **729**(1-2): p. 75-88.

25. Tu, R.H., et al., *Human exposure to the jet fuel, JP-8.* Aviat Space Environ Med, 2004. **75**(1): p. 49-59.

26. Krueger, J.M. and J.A. Majde, *Humoral links between sleep and the immune system: research issues.* Ann N Y Acad Sci, 2003. **992**: p. 9-20.

27. Krueger, J.M. and F. Obal, Jr., *Sleep function.* Front Biosci, 2003. **8**: p. d511-9.

28. Obal, F., Jr. and J.M. Krueger, *Biochemical regulation of non-rapid-eye-movement sleep.* Front Biosci, 2003. **8**: p. d520-50.

29. Krueger, J.M., J.A. Majde, and F. Obal, *Sleep in host defense.* Brain Behav Immun, 2003. **17 Suppl 1**: p. S41-7.

30. Krueger, J.M., et al., *The role of cytokines in physiological sleep regulation.* Ann N Y Acad Sci, 2001. **933**: p. 211-21.

31. Achermann, P. and A.A. Borbely, *Simulation of human sleep: ultradian dynamics of electroencephalographic slow-wave activity.* J Biol Rhythms, 1990. **5**(2): p. 141-57.

Development of a Proton-transfer reaction Ion Trap Mass Spectrometer (PIT-MS) for sensitive online trace gas detection in Life Sciences

M. M. L. Steeghs, S. M. Cristescu, C. Sikkens, and F. J.M. Harren

Molecular and Laser Physics, Institute for Molecules and Materials,
Radboud University Nijmegen, Toernooiveld 1, 6525 E, Nijmegen, the Netherlands
E-mail: f.harren@science.ru.nl, www.tracegasfac.science.ru.nl

Proton-Transfer Reaction Mass Spectrometry (PTR-MS) is a relatively new, yet established technique for measuring trace gases. It is non-invasive, on-line, has a high temporal resolution and provides absolute measurements of trace gas concentrations in a carrier gas. It has shown its potential in various fields, including atmospheric research, food quality control, biological research and medical diagnostics. Conventional systems use a quadrupole mass filter, which has a few drawbacks when measuring complex gas mixtures. Ion trap-based mass spectrometers have a much higher duty cycle and offer the possibility of collision induced dissociation (CID) to identify compounds and distinguish between several isobaric molecules. Here, the early stages of development of a PTR-MS system based on an Ion Trap MS are presented. In the future, the system will be used in medical diagnostics research.

Here the first results of our custom-built Proton-transfer reaction Ion Trap Mass Spectrometer are presented. A PTR-chemical ionization cell has been interfaced with a commercial Thermo Finnigan Polaris Q ion trap system. One of the main stumbling blocks in developing this PIT-MS system is the question whether the ion trapping efficiency can become high enough to be as sensitive as a "conventional" PTR-MS. Here, we show that moderate concentrations of trace gas components can already be measured using our custom-built PIT-MS system (Fig. 1). In addition, trapped ions (Fig. 2) can be kept inside the trap for a prolonged period of time to perform extra analytical steps like Collision Induced Dissociation (CID) or in-trap chemical reactions. Figure 3 shows that it is indeed possible to perform CID to distinguish between two isobaric compounds by looking at the fragmentation spectra after isolation and CID. Different products and different product ratios can be observed for the isobaric compounds propanal and acetone.

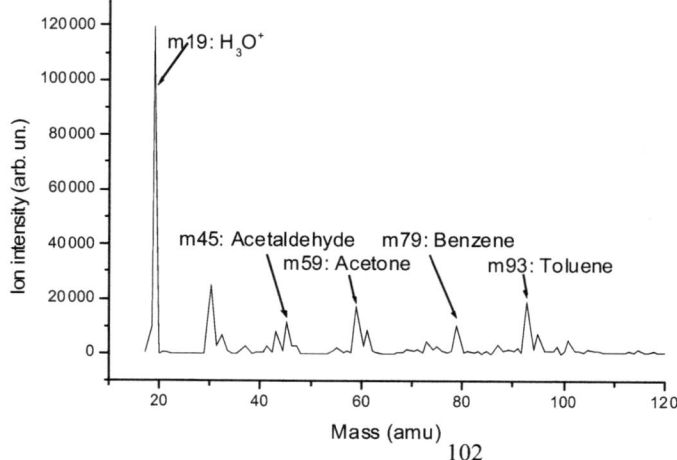

Figure 1:

Mass spectrum of a 0.5 – 1.0 ppmv calibrated mixture of methanol (33), acetaldehyde (mass 45), acetone (59), isoprene (69), benzene (79), toluene (83) and styrene (105)

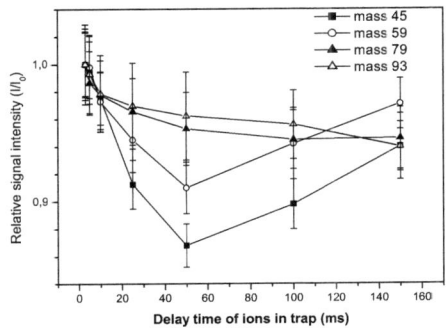

Fig.2. Intensity as a function of the time that the ions are in the trap. The fact that >90 % of all ions are kept inside for a prolonged period of time indicates the basic ability to perform CID

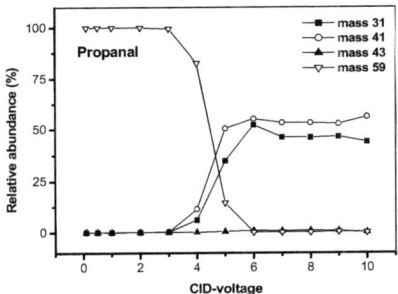 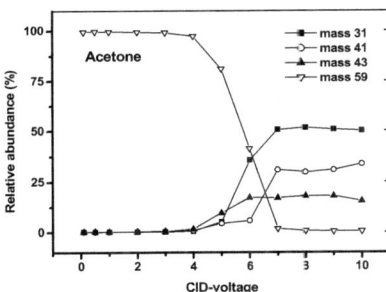

Fig.3 Collision induced dissociation pattern of the two isomers: propanal and acetone. Identification is possible due to different dissociation product ratios

Noise reduction will be a critical step to reach higher sensitivity. Besides, introducing an octopole ion guide into the system will improve the efficiency of the ion transfer from the source to the ion trap. It will also give means to control the translational energy of the ions, which will improve the trapping efficiency of the ion trap mass analyzer. This, in combination with the results shown here, is very promising for the future, in which this system will be implemented in trace gas analysis for biological and medical research, including human breath analysis.

Acknowledgements: this work was financially supported by the Dutch Foundation for Fundamental Research on Matter (FOM) via grant no 02PR2082.

3. Contributed Papers (Posters)

Field intercomparison of PTR-MS and GC measurements of isoprene and its oxidation products during the ECHO campaign 2003

A. Schaub[2,1], W. Grabmer[1], J. Beauchamp[1], A. Hansel[1], C. Spirig[3], B. Mittermaier[2], D. Klemp[2]

[1]*Institute for Ionphysics, University of Innsbruck, A-6020 Innsbruck*

[2]*Institute of Chemistry and Dynamics of the Geosphere: ICG 2: Troposphere, Research Centre Jülich, D-52428 Jülich, Germany*

[3]*Agroscope, Swiss Federal Research Station for Agroecology and Agriculture, Zürich, Switzerland*

Contact: a.schaub@fz-juelich.de

ABSTRACT

During the ECHO field campaign in summer 2003 a PTR-MS was set up for continuous measurements of biogenic VOCs in a mixed deciduous forest. Tower measurements were carried out with a hight time resolution allowing the calculation of VOC fluxed from the forest. A field intercomparison of the data obtained by measurements with an online GC-FID system show a good agreement between the PTR-MS data and of the GC-FID system.

1. Introduction

Isoprene is emitted from many plants and once released into the atmosphere it reacts very fast with the hydroxyl radical. Since this reaction is coupled with formation of ozone it is of great interest to get insight to which extent emitted isoprene is transformed and how large the flux of isoprene is into the atmosphere is. The Eddy covariance technique is a common procedure to calculate fluxes of highly reactive VOCs like isoprene but this technique requires a fast monitoring of the VOC, best in the same time resolution as meteorological measurements have and with a good data quality.

Within the ECHO project (**E**mission and **CH**emical transformation of biogenic volatile **O**rganic compounds) a 36 m high measurement tower was set up in a mixed deciduous forest at the Research Centre of Jülich, Germany. From May to October 2003 a full set of meteorological parameters, photolysis frequencies as well as biogenic VOCs, nitrogen oxides, and ozone were measured continuously. During an intensive campaign in July 2003 a full set of trace gas measurements including radical measurements was carried out to study the combined effect of photochemistry and meteorological parameters within and above a forest stand in detail.

2.

During summer 2003 a PTR-MS run by the Institute of Ionphysics, Innsbruck, was set up at the 36m high tower at the ECHO forest site from July until September 2003. The PTR-MS was operated in high time resolution of 4 Hz for the masses m 41 (methanol), m 69 (isoprene), m 71 (sum of methyl vinyl ketone (MVK) and methacrolein (MACR)) and m135 (sum of monoterpenes).

At the same time an online GC-FID system was measuring isoprene and its oxidation products methyl vinyl ketone and methacrolein, as well as anthropogenic VOCs. The aim of these measurements was to monitor the seasonal behaviour of isoprene mixing ratios and the related pattern of its oxidation products.

The large dataset of ~ 3000 chromatograms and more than 2000 hours of high time resoluted PTR-MS measurements are suited to gain insight in the photochemistry of isoprene emitted from the local forest stand within the scale of seconds as well as in the longterm run.
Both analytical systems took successfully part at an intercomparison experiment prior the field campaign and here we present data of the VOC measurements carried out by PTR-MS and the GC FID.
A field intercomparison between the highly time resoluted PTR-MS measurements and the intergrated sampling time of the GC-FID is a task.

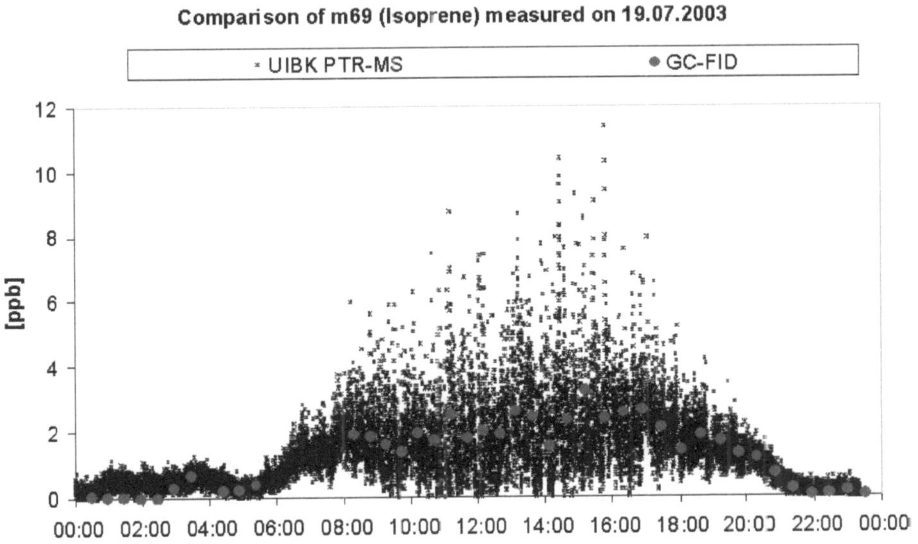

Fig 1: Comparison of m69 (isoprene) mixing ratios measured by PTR-MS (black crosses) with a time resolution of 4 Hz and and GC FID (grey circles) with a time resoultion of 30 min and 5 min intergrated sampling time (Field Campaign of the ECHO project in summer 2003).

Despite the different time resolution, sampling and analyzing techniques the comparison of data shows a good agreement of measured isoprene concentrations. The high time reslution of the PTR-MS shows the abundance of high isoprene fluxes directly above the forest canopy reaching values up to 11.5 ppb which are not captured by the GC-FID.

Based on this agreement the high time resoluted PTR-MS data allows to calculate isoprene fluxes and further data interpretation is planned to combine the knowledge of isoprene fluxes, its photochemical decomposition and meteorological factors.

Measurements of biogenic VOCs in a boreal forest ecosystem

T. Ruuskanen[1], J. Rinne[1], R. Taipale[1], P. Kolari[2], J. Bäck[2], H. Hakola[3], H. Hellén[3], P. Keronen[1], N. Altimir[2], P. Hari[2], M. Kulmala[1]

[1] *University of Helsinki, Department of Physical Sciences, P.O.B. 64, FIN-00014 University of Helsinki, Finland; taina.ruuskanen@helsinki.fi*

[2] *University of Helsinki, Department of Forest Ecology, University of Helsinki, Finland*

[3] *Finnish Meteorological Institute, Climate and Global Change Research, Helsinki, Finland*

ABSTRACT

First measurements of biogenic VOCs in the boreal forest ecosystem in Hyytiälä, South-western Finland, using PTR-MS analyzer were conducted during the summer 2004. Concentrations of the masses of methanol and monoterpene fragment showed distinctive diurnal patterns. Also vertical concentration profiles of these compounds showed highest concentrations near the surface, indicating the forest to be a source of these compounds. Also enclosure measurements conducted on a Scots pine branch showed emission of mass of monoterpene fragment.

1. Introduction

Boreal forests are one of the major vegetation types on Earth, covering about 10 % of land surface. In Eurasia, the boreal zone, or Taiga, extends from Scandinavia through Siberia to the Pacific coast. As the boreal regions are typically sparsely populated, the VOC emissions are dominated by biogenic sources.

Division of Atmospheric Sciences of the University of Helsinki operates two field sites in the boreal ecosystems of Finland (SMEAR I and II). The first one is located in the Eastern side of the Finnish Lapland in the vicinity of Kola Peninsula, and the second in Hyytiälä, South-Western Finland.

The measurements were carried out at the SMEAR II measurement station (Station for Measuring Forest Ecosystem – Atmosphere Relations) in Hyytiälä, Southern Finland ($61°$N, $24°$E, 180 m a.s.l.). The forest around the station is dominated by Scots pine (*Pinus sylvestris* L.) with some Norway spruce (*Picea abies* Karst.), European aspen (*Populus tremula* L.) and birch (*Betula* sp.). The Scots pine forest was planted in 1964.

At the SMEAR II station an extensive set of year round measurements of forest-atmosphere interactions is conducted. These measurements include eddy covariance fluxes of Heat, H_2O, CO_2, O_3 and aerosol particles (Suni et al., 2003), relaxed eddy accumulation measurements of size resolved particle fluxes (Gaman et al., 2004), profile measurements of CO_2, O_3, NO_x

(Vesala et al., 1998). Gas exchange between the atmosphere and plants or soil is measured also with automated and manual chambers (Hari et al., 1999).

Concentrations and emissions of VOCs at this site has previously been studied by methods based on canister and adsorbent sampling with subsequent laboratory analysis (e.g. Rinne et al., 2000, Spanke et al., 2001; Hakola et al., 2003; Hellén et al., 2004). During the summer 2004, measurements of concentrations and emissions of BVOCs have been conducted by a PTR-MS. Some preliminary results are presented in this paper.

2. VOC concentrations and vertical profiles
The VOC concentrations in the air below the forest canopy were measured using a PTR-MS. The sample air was drawn in via two meter long 1/8' Teflon tube. 39 Masses were measured, ranging from 31 to 205 amu, and the integration time for each mass ranged between 1 and 20 seconds.

Vertical profiles were measured using the gas profiling system existing at the SMEAR II station. In this system air is brought down from six heights (4, 8, 17, 34, 50, 69 meters), using Teflon tubing of equal length, with 14 mm i.d. Flow rate in these tubes is 50 l min^{-1} and sample flow to analyzers is taken as a side flow using electronic valves. Each height is measured one minute at a time, leading to six minute cycle. The signal of the valve switching relays was used to trigger the PTR-MS sampling cycle, which measured count rates of 15 masses within less than one minute.

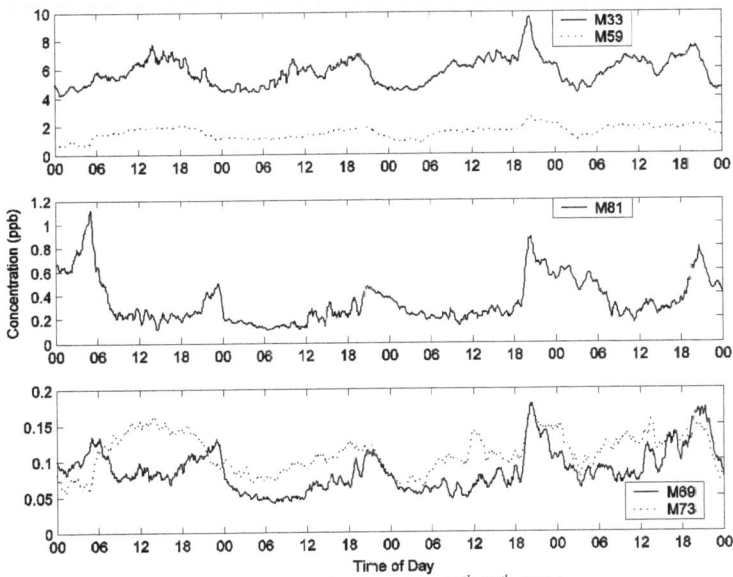

Figure 1: Running mean concentrations of selected masses, July 14th-17th, 2004.

In figure 1 daily cycles of preliminary concentrations of selected masses are shown. Methanol shows at M33, acetone at M59, isoprene at M69, mrthyl-ethyl-ketone (MEK) at M73 and fragment of monoterpenes at M81. M33 Shows maximum concentrations during the late afternoon, whereas M81 shows high concentrations often at night. I figure 2, two daily cycles of count rate profiles of M33 and M81 are shown. Both masses show higher concentrations near the surface indicating a surface source. Accumulation of M81 near the surface can bee clearly seen, as monoterpenes are emitted also during night-time.

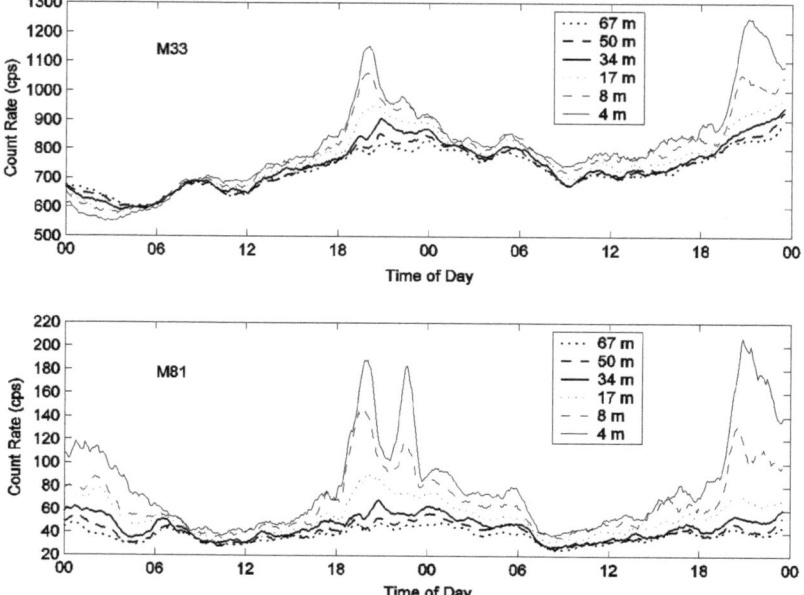

Figure 2: Vertica profiles of running mean count rates of M33 and M81, July 21st-22nd, 2004.

3. Enclosure emission measurements

A scaffolding tower that permits access to the crown of some pines and aspen was used in the emission chamber measurements. Extensive descriptions of the measuring station has been described by Vesala et al. (1998) and the chamber gas exchange set-up by Hari et al. (1999).

The gas-exchange system set-up consisted of chambers, sampling tubing, and analyzers. The closing chambers are open most of the time and are closed intermittently for few minutes. During closure, the gas concentration may show a net increase or decrease depending on the species. For detailed technical description as well as the method of flux calculation see study on ozone fluxes by Altimir et al. (2002).

The measurements were done diurnally in the uppermost part of the canopy, on one-year old shoots. They were installed inside the chamber into a horizontal position, debudded to prevent new growth, and the needles gently bent to form a plane in the same angle as the photo active radiation (PAR) measurements. The temperature was measured by a thermocouple placed within the needles. The emissions are calculations are based on estimates of total needle area.

The PTR-MS was connected to the sample line of the enclosure. Only masses M33 and M81 were measured due to the short closure time of the chamber. In figure 3 a daily cycle of emission on M81 is shown. The highest emissions occur during daytime. However, it seems that the highest emission does not coincide with the highest temperatures, as expected.

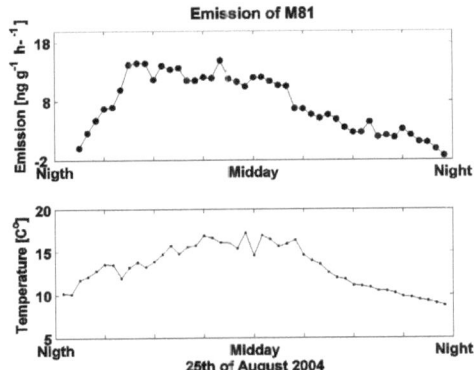

Figure 3: Emission of M81 from an enclosed Scots pine shoot.

References

Altimir, N., Vesala, T., Keronen, P., Kulmala, M. & Hari, P. 2002. Methodology for direct field measurements of ozone flux to foliage with shoot chambers. Atmospheric Enviroment, 36:1, *19-29.*

Gaman, A., Rannik, Ü., Aalto, P., Pohja, T., Siivola, E., Kulmala, M. & Vesala, T., 2004: Relaxed eddy accumulation system for size-resolved aerosol particle flux measurements. J. Atmos. Oceanic Technol., 21, 933-943.

Hakola, H., V. Tarvainen, T. Laurila, V. Hiltunen, H. Hellén & P. Keronen, 2003: Seasonal variation of VOC concentrations above a boreal coniferous forest. Atmos. Environ., 37, 1623-1634.

Hari, P., Keronen, P., Bäck, J., Altimir, N., Linkosalo, T., Pohja, T., Kulmala, M., Vesala, T., 1999. An improvement of the method for calibrating measurements of photosynthetic CO2 flux. Plant, Cell and Environment. 22, 1297-1301.

Hellén, H., H., Hakola, A. Reissell & T.M. Ruuskanen, 2004: Carbonyl compounds in coniferous forest air in Hyytiälä, Southern Finland. Atmos. Chem. Phys., 4, 1771-1780.

Rinne, J., H. Hakola, T. Laurila & Ü. Rannik, 2000: Canopy scale monoterpene emissions of Pinus sylvestris dominated forests. Atmospheric Environment, 34, 1099-1107.

Spanke J., Ü. Rannik, R. Forkel, W. Nigge & T. Hoffmann, 2001: Emission fluxes and atmospheric degradation of monoterpenes above a boreal forest: field measurements and modeling. Tellus, B53, 406-422.

Suni, T., J. Rinne, A. Reissell, N. Altimir, P. Keronen, Ü. Rannik, M. Dal Maso, M. Kulmala & T. Vesala, 2003: Long-term measurements of surface fluxes above a Scots pine forest in Hyytiälä, southern Finland, 1996-2001. Boreal Environment Research, 8, 287-301.

Vesala, T., Haataja, J., Aalto, P., Altimir, N., Buzorius, G., Keronen, P., Lahti, T., Markkanen, T., Mäkelä, J.M., Nikinmaa, E., Palmroth, S., Palva, L., Pohja, T., Pumpanen, J., Rannik, Ü., Siivola, E., Ylitalo, H., Hari, P. and Kulmala, M. 1998. Long-term field measurements of atmosphere-surface interactions in boreal forest combining forest ecology, micrometeorology, aerosol physics and atmospheric chemistry. Trends in Heat, Mass and Momentum Transfer. 4, 17-35.

Characterisation of Italian virgin olive oils by PTR-MS

S. Esposto[1], E. Aprea[2], M. Servili[1], G.F. Montedoro[1], S. M. van Ruth[3]

[1] *Department of Food Science University of Perugia, Via San Costanzo, Perugia, Italy*

[2] *Institut für Ionenphysik, Universität Innsbruck, Technikerstr. 25, 6020 Innsbruck Austria*

[3] *Department of Food Science, Food and Nutritional Sciences, University College, Cork, Ireland*

espostos@unipg.it

ABSTRACT

The Head-Space Analysis (HSA) of olive oil's volatile compounds can be performed using the Proton Transfer Reaction (PTR) Mass Spectrometry (MS).

This paper reports the application of this instrument to classify the virgin olive oils according to the variety, origin area and the olive's ripening stage.

1.INTRODUCTION

Because of the role they play in so many different fields, the volatile compounds are widely investigated and many techniques have been proposed to analyse and monitor them. The sensitivity of some of these methods available today is very high, but normally the price to be paid is high in terms of number of drawbacks in the measurements, e.g., in the sample preparation, time needed, calibration standards, data interpretation and necessity of highly trained people.

Volatile compounds determine the flavour of Virgin Olive Oil (VOO). The substances responsible of olive oil aroma belong to the following chemical classes: esters, aldehydes, ketones, aliphatic alcohols, hydrocarbons, oxygenated terpenes, fatty acids, furan and tiophene derivatives.

VOO sensory quality is currently determined by the European Union Regulation or by the International Olive Oil Council trade standards (1, 2). Both official methods carry out the sensory evaluation by using panels of trained assessors. However, panel test is a costly and slow procedure that is not always at the disposal of small producers or cooperative societies; only large retailers and sellers may be able to afford such tests.

The alternative solution to the panel test for virgin olive oil recognition according to different origin areas, varieties or olive ripening stages is based on the quantification of volatile compounds, which can be evaluated by PTR-MS (3). This technique has been successfully used in many fields of analytical chemistry.

The present study concerned the identification of differences among several Italian VOOs by PTR-MS headspace analysis.

2.MATERIALS AND METHODS
Materials.

To study the differences in volatile profiles among VOO, samples from several Italian towns (Perugia, Roma, Pescara, Cosenza, Andria, Catanzaro), varieties (Moraiolo, Canino, Carolea, Coratina, Frantoio) at three ripening stages of the fruit were analysed by PTR-MS. The VOO were obtained in a pilot plant of the Department of Food Science (University of Perugia).

Methods.

Headspace analysis. The headspace measurements were performed by Proton Transfer Reaction Mass Spectrometry (PTR-MS). This method represents a relatively new approach of chemical ionisation mass spectrometry for the measurement of volatile organic compounds (4). The headspace sampling was carried out according to Biasioli et al. (5).

Statistical Analysis.

The PTR-MS data were subjected to Principal Component Analysis (PCA) using the Software WinDas (6).

3.RESULTS AND DISCUSSION
Fig. 1 shows the PCA plot of the 1st and the 2nd PCs. Along the 1st PC, the VOOs samples were separated in three clusters according to the different varieties, whereas in the individual clusters the three VOOs' ripening stages, were discriminated along the 2nd PC.

Fig. 2 demonstrates the loadings of individual ions on PC2. It preserves the spectra shape and gives an indication of the masses responsible for the differences and consequently of the chemical compounds related. A list of possible attributions is given (7).

Fig. 3 presents the PCA plot of the 1st and 2nd PC showing the discrimination of the different VOOs origin regions.

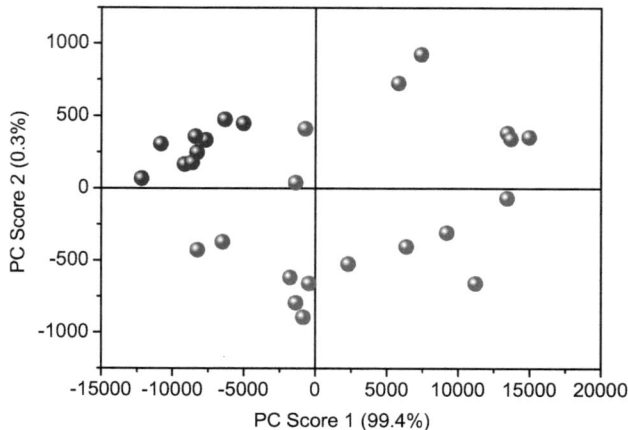

Figure 1. PCA plot of VOOs of three Italian olive's cultivars from the same town and three different ripening stages. Legend: Varieties. Cor: Coratina Cv., Fra: Frantoio Cv., Car: Carolea Cv. Town. PE: Pescara. Maturation states I, II, III: the 1st, the 2nd and the 3rd ripening stage. Each * is an individual repetition.

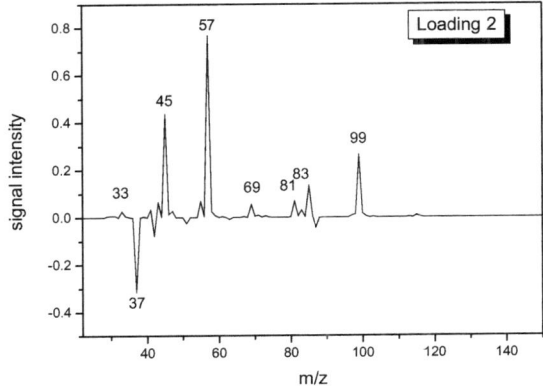

Possible Attributions (6):

33 Methanol

37 Water Cluster $(H_2O) \cdot H_3O^+$

45 Acetaldehyde

57 Alcohol Fragment

69 Aldehydes Fragment

83 & 55 Hexanal

99 & 81 Hexenals

Figure 2. Loading of ions on PC2

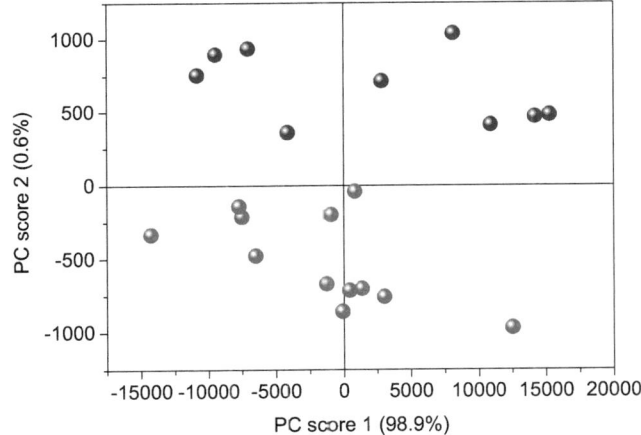

Figure 3. PCA plot of VOOs of two Italian olive's cultivars from several regions, at the same ripening stage. Legend: Varieties. Fra: Frantoio Cv. and Car: Carolea Cv. Towns. PG: Perugia; RM: Roma; PE: Pescara; CS: Cosenza; And: Andria. Maturation state. II: the 2nd ripening stage. Each * is an individual repetition.

4. CONCLUSION

The PTR-MS technique is an important tool for on-line and high sensitivity monitoring of volatile compounds. In this work we showed the possibility to use the PTR-MS fingerprinting to discriminate the VOOs' samples in order to their origin areas, cultivars and olives' maturation.

REFERENCES

1) E.C. Off. J. Eur. Communities. Regulations 2568/91 and 1513/01.

2) IOOC: International Olive Oil Council. 1996. COI/T.15/NC no 3-25. Resolution RES-3/75-IV/96 (revised June 2003).

3) Yeretzian, C., A. Jordan, H. Brevard, W. Lindinger. Flavour Release. ACS, USA, 2000, p.58 (1995).

4) Lindinger, W., A. Hansel, A. Jordan. (1998). Int. J. Mass Spect. Ion Proc. 173: 191-241.

5) Biasioli, F., F. Gasperi, E. Aprea, L. Colato, E. Boscaini, T.D. Märk. (2003) Int. J. Mass Spect. 223-224: 343-353.

6) WinDas. Kensley, E. K., Wiley: Chichester, U.K. (1998).

7) Buhr K., van Ruth S.M., Delahunty C. (2002). Int. J. Mass Spect. 221: 1-7.

On-line Analysis of Organic Compounds in Diesel Exhaust Using Proton Transfer Reaction Mass Spectrometry

M.L. Alexander, B.T. Tom Jobson, G. Maupin and G. Muntean

Pacific Northwest National Laboratory, 3335 Q Ave., Richland WA 99352
michael.alexander@pnl.gov

ABSTRACT

The key goal result of this research is to understand the complex organic mass spectra produced by Diesel exhaust at a suffient level of detail so the PTR-MS can be used as a tool for answering critical questions presented by current engine catalysis research and development efforts. PTR-MS mass scans were made to 210 m/z at different engine loadings. Initial measurements were made by dilution of exhaust from a diesel generator followed by sampling by the PTR-MS. Further measurements were made inside the catalyst of a Volkswagen Passat at several locations using a spatially resolved mass spectrometry techniques, (SPACI-MS). The mass spectra were complex and showed a strong pattern of 14n+1 peaks with a relative abundance similar to that obtained from electron impact ionization of alkanes. Further experiments verified that H_3O^+ proton transfer reaction with n-alkanes under the conditions of the drift resulted in fragmentation patterns nearly identical to those of electron impact. Alkane fragmentation likely simplified the upper end of the mass spectrum, and reduced the mass interference with isobaric aromatics. Tentative identification of several aromatic species and light alcohols and aldehydes was made. It is shown that concentrations and relative abundances of hydrocarbons changed as a function of engine load Concentrations of aldehydes and ketones dominated the exhaust emissions. The relative abundance of alkanes, aromatics, aldehydes, and alcohols as determined by PTR-MS was broadly consistent with literature publications using gas chromatography analysis. About 75% of the PTR organic ion signal could be assigned.

1. Introduction

Diesel and gasoline engines are important sources of nitrogen oxides, hydrocarbons, and fine particulate matter in the urban atmosphere and have major impacts on air quality and human health. In the US new regulations are being phased in over the next several years to reduce particle emissions from diesel engines by 90% by the year 2007. Particle mass emissions rates and composition from diesel engines are determined by a number of factors including engine technology, engine load, air to fuel ratio, fuel parameters, exhaust temperature and driving conditions. To reduce particle emissions, diesel exhaust aftertreatment devices such as the soot filter and catalyst examined in the work presented here are being developed. Complex mixture analysis by this technique can not provide resolution of individual constituents to the same degree as gas chromatography based analyses, but may be the best quantitative technique to date for providing on-line, high time resolution information on selected organic species such as aromatics and light oxygenates.

2. Experimental

For the first studies, Diesel exhaust from a 4 kilo-watt diesel generator was monitored. Exhaust gas passed through a cordierite soot filter. The diluters and stainless steel transfer lines from the soot filter were heat traced to 200 °C and sampling lines on the PTR-MS were controlled at 80 °C. Mass scans were made on diesel exhaust from 20 to 210 amu with 0.5 second dwell times. Thus it took about 1.6 minutes to perform a mass scan. Zero air provided by a heated catalyst was used to monitor of instrument background to check for possible contamination of instrument sampling lines and the drift tube due to high hydrocarbon concentrations in the exhaust. Average background count rates were subtracted from the diesel exhaust data. No major contamination of the instrument was observed.

The second phase of the work involved sampling a catalyst connected to a Volkswagon Passat with a Diesel engine. This catalyst is designed to reduce hydrocarbon emissions by oxidation to carbon dioxide and water. Several locations upstream, downstream and inside the catalyst were monitored using Spaci-MS with the PTR-MS as the mass spectrometer. In SPACI-MS mutiple capillaries, sampling discrete spatial locations are connected to a rotary valve also connected to the PTR-MS. Rapid switching of the valve allows different locations in the catalyst to be sampled under various engine conditions. The PTR-MS was set to monitor only a few mass values to maximize time reponse. The capillaries, rotary valve and PTR-MS inlets were temperature controlled at 80 °C.

In addition to the exhaust sampling, the ion products of H_3O^+ reactions with C_8-C_{16} n-alkanes and C_8-C_{13} 1-alkenes in controlled laboratory experiments with pure compunds were determined to aid the interpretation of the exhaust gas mass spectrum. These experiments used the same ion drift conditions as in the exhaust sampling.

3. Results and Summary

The PTR mass spectrum for the generator engine under idle conditions is shown in Figure 1. Background count rates have been subtracted, removing major ions produced in the ion source such as m/z 21, 32, 37, and 55. The mass spectrum is complex, with a peak at every mass, even numbered masses dominating. Count rates for most masses increased as engine load decreased. This is in contrast to the soot concentration which increased with engine load. An explanation of the mass assignments with respect to the major groups and species, alkanes, alkenes, alkynes, aldehydes, ketones and aromatics is beyond the scope of this abstract and will be given at the presentation. listed in Table 2 is given below. One key feature

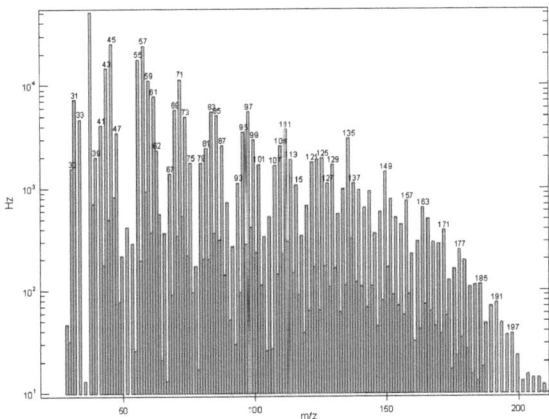

of the interpretation of the mass spectrum is understanding the fragmentation of alkanes in the PTR-MS.

Figure 1. PTR mass scan of diesel engine generator exhaust operating at 0% load. Ion signal intensity in counts per second (Hz) are plotted against mass to charge ratio (m/z). Background count rates have been subtracted from the spectrum. The off scale peak at m/z = 37 is the first hydronium ion water cluster $H_3O^+(H_2O)$.

Figure 2 compares the relative abundance of the 14n+1 masses for the n-alkanes, a reference electron impact mass spectrum of tridecane to show the similarity to the PTR fragmentation pattern, and the exhaust samples at different engine loads. The relative abundance of the 14n+1 peaks in the exhaust samples diminished more gradually at higher masses than the n-alkane mass spectra. One reason for this might be the presence of branched alkanes. By analogy with the n-alkanes and the similarity of their product fragments to those observed from electron impact, the branched alkanes also likely react with H_3O^+ fragmenting to similar products. The relative intensity of the 14n+1 mass fragments for isoalkanes ionized by electron impact varies depending on molecular structure. The relative large ion signals at m/z 99, 113, 127 observed in the exhaust samples may be due to branched alkanes which are known to be a significant component of diesel fuel. We conclude that the PTR-MS was responding to C_{10} to C_{20} alkanes and isoalkanes present in diesel exhaust, giving rise to

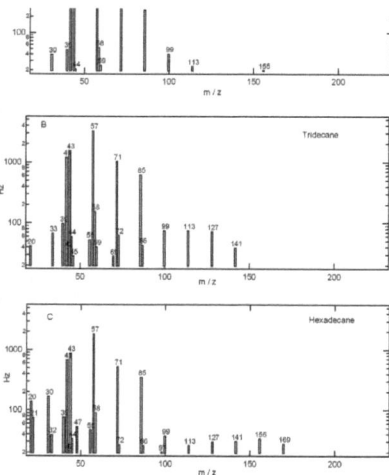

significant signals at masses 43, 57, 71, 85, 99, and 113. The sum of these mass signals could be used to estimate the total concentration of C_{10}-C_{20} alkanes in the exhaust.

Figure 2. Ion products of the reaction of H_3O^+ with a series of n-alkanes a) undecane b) tridecane c) hexadecane.

The results of the spatially resolved PTR-MS in a catalyst are still being evaluated. Figure 4 shows a sample of the preliminary results.

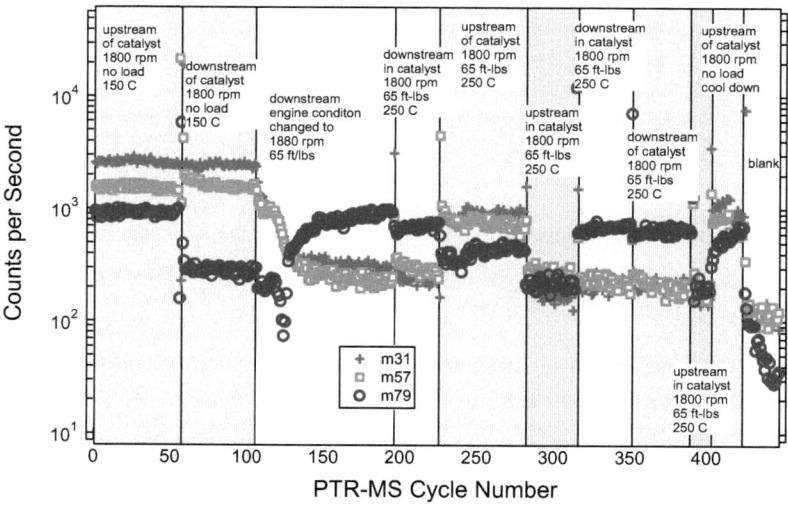

Figure 4 Preliminary Data from SPACI-PTR-MS measurements in Diesel catalyst for formaldehyde (m31), an alkane fragment ($C_4H_9^+$ m57) and benzene (m79)

While the data from m31 and m57 show the expected behavior, little catalyst effect at low temperature with full oxidation at high temperature, the behavior of benzene was erratic and often the reverse of expected. The other clear result is that at high operating temperature when the catalyst is most efficient, only the first cm or less is needed for complete oxidation of organics. Further results and revised conclusions will be presented.

Measurement of methanol emissions of tobacco seedlings by PTR-MS

M. Müsch[1], D. Abanda[2], W. Schwab[2], J. Tschiersch[1]

[1]GSF – National Research Center for Environment and Health, Institute of Radiation Protection, Ingolstaedter Landstr. 1, D-85764 Neuherberg, Germany, (muesch@gsf.de)

[2] Technical University of Munich, Biomolecular Food Technology, Lise-Meitner-Str. 34, D-85354 Freising, Germany

ABSTRACT

In this work the influence of Pink Pigmented Facultative Methylotrophic bacteria (PPFM) on the methanol emissions of tobacco seedlings is analysed. For its low detection limit and rapid response a Proton-Transfer-Reaction Mass Spectrometer (PTR-MS) was chosen for the measurements. The methanol produced by abacterial tobacco seedlings could be detected, while no methanol emissions could be observed for the plants inoculated with PPFM which proofs the effect of PPFM. Apart from methanol, several other Volatile Organic Compounds (VOCs) were detected. For a period of about one week the time course of the VOC concentrations were measured.

1. Introduction

Pink Pigmented Facultative Methylobacterium (PPFM) is the most abundant methylotrophic bacteria and can be found in the phyllosphere of more than 50 plant species as well as in the rhizosphere [1]. These bacteria live in symbiosis with plants. PPFM is able to use the methanol formed as a waste product by metabolic processes in the plants [2]. This step is very difficult to prove by analytical means like GC-MS, because the concentrations of methanol are relatively low if there is only a small group of seedlings in a vessel. The task was to find evidence by PTR-MS that plants inoculated with PPFM do not emit methanol while inert plants do. The time course of the emitted methanol was of special interest, but the evolution of other VOC concentrations was also determined to test the measurement method and to obtain more information on the total VOC production by the seedlings.

2. Preparations and measuring procedure

For the experiments a Proton-Transfer-Reaction Mass Spectrometer (PTR-MS) built by Ionicon (Austria) was used. In this instrument VOCs are protonated by H3O+-ions in a drift chamber and subsequently detected by a quadrupole mass spectrometer. The main components of air do not take part in this reaction. Thus VOCs can be measured online with a low detection limit [3].

After some pre-experiments tobacco seeds were chosen for the experiments because methanol emissions by abacterial seedlings could be measured and the seedlings develop fast enough. For the measurements about 75 tobacco seeds with water were given on a germination filter in a close glass vessel with two opposite screw connections, each sealed with a septum (Fig. 1). Needles were introduced into both of them, one with a PTFE tube of about 50 cm length leading directly into the inlet of the PTR-MS, the other one serving for pressure equalization.

In pre-experiments it could be seen already that the attenuation of the sample gas in the glass with the room air is negligible because methanol is formed very quickly and in large amounts by the seedlings.

PTR-MS  room air

Fig. 1: Glass vessel with seedlings and screw connections

Five glasses were prepared with sterile tobacco seeds, another five vessels with seeds inoculated with PPFM. The experiments were carried out at room temperature. Storage temperatures of the samples were in the range of about 20°C with an exposure of 12 h to infrared light each day.

Measurements were carried out daily for about one week. Each sample was measured twice every day. Between the samples the indoor room air was measured as reference gas. The measurement program was developed in pre-experiments. For methanol (protonated mass 33) a measuring time of 2 s was chosen, for the other protonated masses (34, 37, 41, 42, 43, 45, 49, 50, 51, 55, 59, 61, 62, 63, 73, 75, 83, 85, 95) 50 ms each.

3. Results

At the beginning of the experiment the seedlings were 7 days old and had already formed their first two leaves. The plants inoculated with PPFM were obviously better developed with a greater size and bigger leaves.

In none of the measurements of the inoculated seedlings the methanol concentrations exceeded the background level, i.e. the concentration in the room air. Exemplary data is given for the measurement on day 4 in Fig. 2 (section II). The concentration of methanol seemed to be even lower in the vessel than in indoor air. On the other hand methanol could be found in every abacterial sample (I). The level was clearly higher than in the reference gas.

One sterile and one inoculated sample were measured for a longer time of about 20 min (Ia, IIa). For the sterile sample a high constant methanol concentration was observed while the inoculated sample did not show any increase. From this observation the conclusion can be drawn that the vessels were close. Furthermore the sample air was not diluted remarkably by the air flowing in through the second needle.

The glass container of the abacterial sample used for the long-time test was also opened for 10 minutes and then closed again. About half an hour later, the level of methanol concentration had already reached again the level before the glass had been opened (IIb).

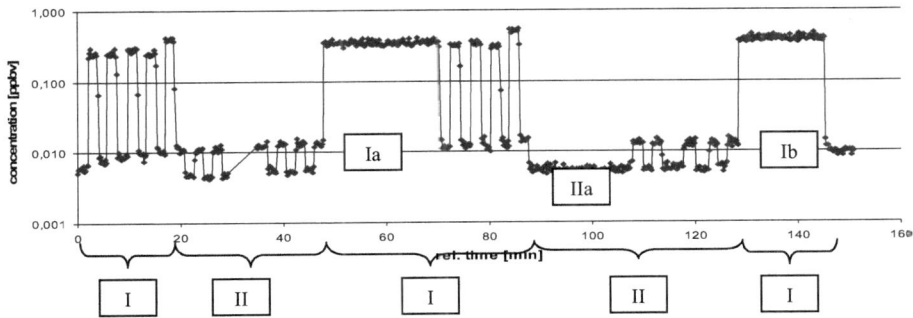

Fig. 2: Measurement data from day 4 (section I: sterile samples; Ia: long-time measurement; Ib: long-time measurement after opening and re-closing the vessel; II: samples inoculated with PPFM; IIa: long-time measurement)

In the course of the experiment it could be observed that the seedlings kept on growing whereat the tobacco plants with the PPFM bacteria grew better, i.e. they built more biomass, were taller and had bigger leaves (Fig. 3).

Fig. 3: Photograph of tobacco seedlings. Left: sterile sample; right: sample inoculated with PPFM; day 8 of experiment (seedlings 15 days old).

During the experiment the average concentrations of methanol in the gas phase above the abacterial seedlings decreased (Fig. 4). Each average value was formed by 10 measurement values (2 measurements at each of the 5 samples). The average background was subtracted except for methanol with PPFM, because in this case the concentration in the room air was slightly higher than in the sample gas. A behavior corresponding to the methanol concentration could be observed for a number of other substances as well, e.g. for M42. In the case of acetone (M59), only for the inoculated seedlings a clear decrease could be found. Some of the measured substances have not been identified yet, e.g. M42. One VOC, M45, was emitted only by the plants inoculated with PPFM.

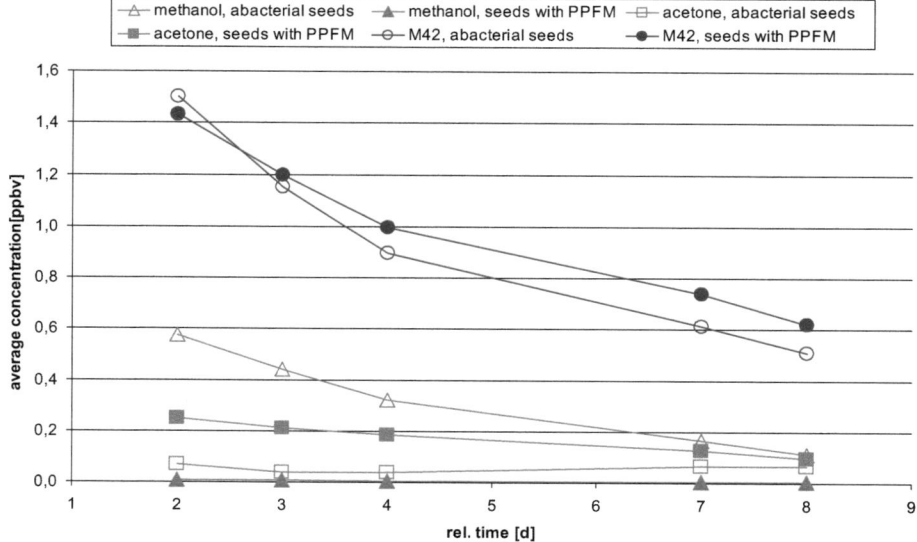

Fig. 4: Average concentrations of various VOCs by inert and inoculated tobacco seedlings

4. Conclusions

The measurement method was applied successfully. PTR-MS could be used as well for a number of similar problems. The setup for the experiment is simple and short measuring times can be used.

In our experiments it could be clearly shown that seedlings without PPFM bacteria emit methanol whereas inoculated plants do not. The emission rate depends on the age of the seedling. In this respect it would be interesting to investigate the VOC emissions of a newly planted seed sample to see when the methanol production is the highest in the development of the seedling. The identification of the other measured VOCs would be important, also in regard to an estimation of the total emission rates of VOCs by vegetation into the atmosphere. Furthermore experiments with other plant seedlings would be useful in order to obtain more general information on the symbiosis between plants and bacteria.

References

1. Trotsenko, Yu. A., Ivanova, E. G., Doronina, N. V., Aerobic Methylotrophic Bacteria as Phytosymbionts, Microbiology, 2001, vol. 70, no. 6, pp. 623-632.
2. Romanovskaya, V. A., Stolyar, S. M., Malashenko, Yu. R., Dodatko, T.N., The ways of Plant Colonization by Methylobacterium Strains and Properties of These Bacteria, 2001, Microbiology, vol. 70, no. 2, pp. 221-227.
3. Hansel, A., Jordan, A., Holzinger, R., Prazeller, P., Vogel, W., Lindinger, W., Proton transfer

   reaction mass spectrometry: on-line trace gas analysis at the ppb level, International Journal

   of Mass Spectrometry and Ion Processes, 1995, vol. 149/150, pp. 609-619

Investigating the effect of ozone treatment on the microbial flora of pork

D. Jaksch[1,2], **T. Mikoviny**[1], **E. Hartungen**[1], **N. J. Mason**[2], **T.D. Märk**[1,3]

[1]*Institut fuer Ionenphysik, Technikerstr. 25, A-6020 Innsbruck, Austria*

[2]*Department of Physics and Astronomy, The Open University, Walton Hall, Milton Keynes MK 7 6AA, UK (d.jaksch@open.ac.uk)*

[3]*Department of Plasmaphysics, Comenius University, 84248 Bratislava, Slovak Republic*

ABSTRACT

We report the results of treating commercial samples of pork meat with ozone in order to determine whether such treatment reduces microbial growth and hence extends the shelf lifetime of such products. The technique of PTR-MS was used to study volatile emissions with the signal detected at mass 63 (assumed to be a measure for dimethylsulphide) being used as a diagnostic of bacterial activity. Such a signal was found to strongly increase with time for an untreated meat sample whereas ozone treated meat samples showed much reduced emissions - suggesting that the microbial activity had been greatly suppressed by ozone treatment. An independent analysis however revealed that microbial counts were very high, independent of the treatment.

1. Introduction

An estimated 30% of fresh produce is lost by microbial spoilage from the time of harvest, through handling, storage, processing, transportation, shelving and delivery to the consumer (web1). To preserve food pathogens need to be destroyed or inactivated and the non-pathogenic microorganisms and enzymes responsible for food spoilage need to be eliminated or at least reduced (web2). Several techniques to extend food's shelf-life have been developed over the years for example heating, drying, irradiation and treatment with ozone. All these methods have their advantages, drawbacks and limitations depending on the type of the food, the kind of microorganisms, national regulations and the public demands (like unaltered taste, aroma, colour and vitamin content with no chemical residues after treatment). The use of ozone to treatment of the food using ozone gas meets these requirements quite well.

Ozone is a strong oxidant that kills many microbacterial organisms without leaving any toxic by-products or residues (web3, web4). Ozone has also been used for many years to treat pathogens such as bacteria and algae in water for applications such as drinking water supplies (web4) - ozone should therefore be a useful agent for the destruction of pathogens which are active in microbial spoilage of meat. Despite these advantages the use of ozone in the food industry has not been exploited as extensively since ozone must be manufactured on-site and until recently ozone generators were bulky and expensive (web5). However new developments in the design of small scale in situ ozone generators (using either UV lamps or electrical discharges) now make it practical to develop ozone treatment for food preservation on a commercial scale.

To date there have been only a few studies to quantify the ozone concentrations needed to ameliorate microbial spoilage. To fill the gap in this knowledge the aim of this study is to investigate the influence of ozone on microbial spoilage using the technique of PTR-MS to analyse VOC emissions derived from microbial spoilage.

In an earlier study (Mayr *et al.*, 2003), it has been shown that the emission of some specific VOCs are characteristic of microbacterial activity, for example emissions detected at mass 63 are statistically significant strongly correlated to the aerobic counts, to the counts of *Pseudomonas* spp., Enterobacteriaceae and *Enterococcus* spp. respectively in air-packaged meat (pork and beef) hence monitoring VOC emissions from the food provides a direct methodology for assessing bacterial activity. In contrast to technique of counting bacteria (requiring the incubation period of 1-3 days) detection of VOCs may be performed online and with rapid sampling rates. The use of a PTR-MS system provides an on-line measurement of VOCs with concentrations as low as a few parts per trillion in volume (pptv) and thus may detect bacterial activity when it is just beginning. Another advantage of PTR-MS is that the samples containing the volatile compounds do not need any preparation (pre-sampling, pre-concentration or sample dehydration) before being admitted to the PTR-MS. Thus some problems inherent to sampling in alternative methods used for VOC detection (e.g.: gas chromatography) are avoided, since the food itself is not disturbed and the measured mass spectral profiles closely reflect genuine headspace distributions (Yeretzian *et al.*, 2002).

2. Material and methods

Two sets of measurements were performed six months apart. In each case retailed pork cutlets that were air packaged in an oxygen-permeable polyethylene film were bought in a supermarket in Innsbruck on the day when the respective measurements were started. Their expiry date was listed as two (first set of measurements, i.e. Experiment (Exp) 1 and 2) and three (second set, i.e. Exp 3) days after purchase.

Pieces of about the same shape (approx. 35x50x10mm), weight and consistency were cut out of a single cutlet for Exp 1-3, respectively. Each sample was placed into a glass flask (volume V=300ml) with a metal cover containing two gas inlets through which gas could be passed over the meat sample. The control samples were flushed with oxygen/air the others exposed to different concentrations of ozone (see table 1) for ten minutes. All the treated flasks were subsequently stored under identical conditions at room temperature. Measurements of the VOC emissions from the samples were then made at regular intervals over the period of several hours (see table 1). Headspace air was drawn through a heated teflon transfer line into the PTR-MS system for on-line VOC analysis. The mass spectrometric data being collected over a range of masses (m) with m/z=20–150amu, where z is the charge of the measured ions (in our case z=1), different m being characteristic of different VOCs - in turn a monitor of different microbacterial processes.

3. Results and Discussion

The effect of ozone treatment on the pork's decay behaviour was monitored through the observation of the concentration detected at mass 63 assumed to be dimethylsulphide (DMS) as this signal has been shown to have the largest correlation (up to 99%) with the bacterial contamination of meat (Mayr et al., 2003).

Figure 1 shows the results of the Exp 1-3. After a certain time lag the DMS signal detected from the oxygen treated sample in Exp 1 strongly increased with time whereas the low dose ozone treated sample showed only a slight increase, and the signal of the high dose treated pork piece remained almost constant. The same emission behaviour was found for the first part (t=0-30h) of Exp 2. However, the oxygen treated sample was exposed to a high dose of ozone at t=30h and the DMS concentration was found to strongly decrease, indeed, it took about 9 hours until the initial concentration was reached again. Comparing the results of Exp 1 and 2 one can see the strong influence of the additional ozone treatment on the emissions of the oxygen treated pork samples. The DMS concentrations of the both oxygen treated samples in Exp 1 and 2 were similar before the exposure to ozone at t=30h. In Exp 1 signals from the non ozone treated sample reached a concentration of 1.3×10^3 ppb at the end of the measurements (t=46h, Fig. 1) whereas in Exp 2 the DMS concentration of the ozone treated sample was only 90ppb at t=46h. The online monitoring in Exp 2 was concluded at t=100h (not shown in Fig.1). The highest DMS signal (with a concentration of 300ppb) of ozone treated meat was reached at t=68h and stayed constant for six hours and was much lower than the highest measured DMS signals from the non ozone treated samples in Exp 1. The trends seen in the first two experiments were confirmed by the results of Exp 3. The DMS signal of the untreated and oxygen treated samples strongly increased with time – less strongly for the oxygen treated pieces. The oxygen treated samples were exposed to ozone after 42 hours with a subsequent decrease in the detected DMS signal and the concentrations remained low until the end of the experiment. The highest ozone exposure resulted in the detected DMS signal showing nearly no increase during the whole measurement time.

The microbiological analysis revealed that microbial counts were very high, independent of the treatment (Jaksch et al., in print).

		TREATMENT				
		untreated	O_2	O_2+high[a] O_3	low[b] O_3	high[a] O_3
Exp 1	no of samples	-	1	-	1	1
	analysis period (h)	-	47	-	47	47
Exp 2	no of samples	-	-	1	1	1
	analysis period (h)	-	-	100 (30)[c]	30	30
Exp 3	no of samples	2[d]	2[d]	2[d]	2[d]	2[d]
	analysis period (h)	46	46	49 (42)[c]	46	46

[a] 1000ppm [b] 100ppm [c] Ozone treatment after (x) hours [d] microbiological analyzed

Table 1: Samples were cut out of a single pork cutlet for Experiment (Exp) 1-3 respectively, treated in the different ways for 10 min and stored under identical conditions at room temperature. The emissions were regularly measured over the given time period. A microbiological analysis was performed for the samples of Exp 3 at the end of the analysis period.

4. Conclusion

In the present work we have shown the strong effect of ozone exposure on pork cutlet's emissions, which have been found earlier to be highly correlated to the bacterial contamination, suggesting its usefulness as a remedial action for microbial spoilage to extend food shelf life. Even a later treatment with ozone strongly delayed the bacterial activity. The reduction of VOCs on one hand and the high microbial counts on the other hand indicate that the treatments applied in this study were effective to inhibit and thus reduce physiological activities, but are not necessary effective enough to produce a lethal effect on microorganisms present in meat. Further studies are needed to optimize the use of ozone in order to reduce microbial spoilage of meat.

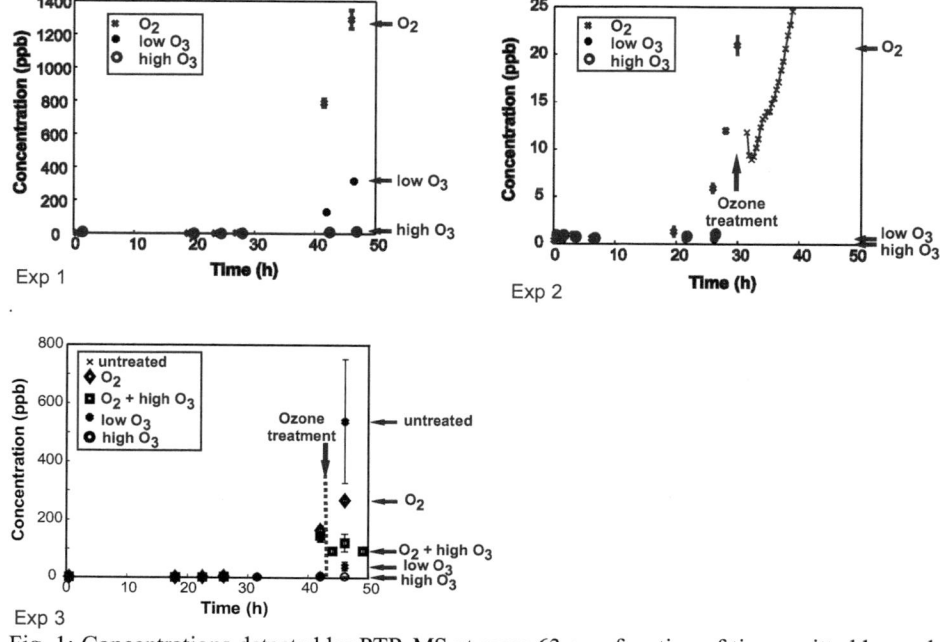

Fig. 1: Concentrations detected by PTR-MS at mass 63 as a function of time emitted by pork samples that were differently treated for 10 min. prior to the first measurement at time t=0. These results suggest ozone significantly retards microbial spoilage.

5. References

Lindinger W., Hansel A., Jordan A., 1998. On-line monitoring of volatile organic compounds at pptv levels by means of Proton-Transfer-Reaction Mass-Spectrometry (PTR-MS): Medical applications, food control and environmental research, *Int. J. Mass Spectrom. Ion Processes*, **173**: 191-241.

Mayr D., Margesin R., Klingsbichel E., Hartungen E., Jenewein D., Schinner F., Märk T.D., 2003. Rapid Detection of Meat Spoilage by Measuring Volatile Organic Compounds (VOCs) Using Proton-Transfer-Reaction Mass-Spectrometry (PTR-MS), *Appl. Environ. Microbiol.*, **69**: 4697-4705.

Jaksch D., Margesin R., Mikoviny T., T.D. Skalny, Hartungen E., Schinner F., Mason N.J., Märk T.D., The effect of ozone treatment on the microbial contamination of pork meat measured by detecting the emissions using PTR-MS and by enumeration of microorganisms, *Int. J. Mass Spectrom.*, in print.

Web1: http://www.zentox.com/Ozofood.pdf

Web2: http://ip.cals.cornell.edu/courses/intag402/readings/processingtech.pdf

Web3: http://www.appliedozone.com

Web4: http://www.ecosensors.com/pg4_1applozonetnap_101.html

Web5: http://aggie-horticulture.tamu.edu/extension/newsletters/foodproc/dec01/rrart1.html

Yeretzian C., Jordan A., Badoud R., Lindinger W., 2002. From the green bean to the cup of coffee: investigating coffee roasting by on-line monitoring of volatiles, *Eur. Food Res. Technol.*, **214**: 92-104.

Methanol emissions from a grassland system

Aurelia Brunner[1], Christof Ammann[1], Albrecht Neftel[1], Christoph Spirig[1], Johannes Stähelin[2]

[1]Agroscope FAL Reckenholz, Swiss Federal Research Station for Agroecology and Agriculture, CH-8046-Zürich, aurelia.brunner@fal.admin.ch

[2]Institut für Atmosphäre und Klima, IAC, ETH-Hönggerberg, CH-8093 Zürich

Volatile organic compounds (VOCs) are part of the plant metabolism. They are produced e.g. to attract or repel insects or as wound protection means. Once released, they play an important role in tropospheric chemistry. Reactions of VOCs with photochemical oxidants lead to the formation of ozone, aerosol particles and other compounds that can be toxic for men and plants. Relative to the number of investigations about VOC emissions from forest only few studies on grassland exist.

The aim of this study is to characterize the VOC emission from grassland systems. From June 7th until September 20th 2004 a field experiment was carried out at the Swiss CarboEurope grassland site Oensingen (coordinates: 7°44'E, 47°17'N, altitude: 450 m a.s.l., rainfall: 1109 mm, mean annual temperature: 9°C). The field has a size of 100x140m and is divided in an intensively and an extensively managed part. Fluxes of selected VOCs were determined by two different approaches: eddy covariance (EC) and dynamic chambers. For both techniques the VOC concentrations were measured by a proton-transfer-reaction mass spectrometer (PTR-MS.

Methanol was the dominating VOC species emitted from the grassland. Average methanol concentrations above the canopy were in the order of 10 ppb. After mowing the concentration increased over 100 ppb and coincided with the maximum observed flux of 7 mg/(m^2 h). The flux then decreased exponentially within a few hours to a flux of 0.5 mg/(m2h). Methanol dominated the VOC fluxes both on the intensively and the extensively managed field. The dynamic chamber measurements are compared to the parallel EC flux measurements.

Influence of oral processing of strawberry flavoured custards on in-nose flavour concentrations measured by PTR-MS

Eugenio Aprea[1,2,3], Franco Biasioli[2], Flavia Gasperi[2], Tilmann Märk[3], Saskia van Ruth[1]

[1]*University College Cork, Department of Food and Nutritional Sciences, Western Road, Cork, Ireland, email: s.vanruth@ucc.ie*

[2]*Instituto Agrario di S. Michele a/A, S. Michele, Via E. Mach 2, 38010, Italy*

[3]*Institut für Ionenphysik, Universität Innsbruck, Technickerstr. 25, A-6020 Innsbruck, Austria*

ABSTRACT
The interaction of oral processing protocol and food texture on *in vivo* flavour release was examined using in-nose PTR-MS analysis. Custards with different concentrations carboxymethyl cellulose (CMC) were compared, as well as free chewing and an imposed oral processing protocol. Significant effects of type of compound, custard texture, oral processing protocol and subjects were observed.

1. Introduction
Sensory perception of the flavour and texture of food products depends on the composition and the structure of the food systems. Variables such as hardness, waterholding capacity or microstructure have been shown to affect the perception of flavour. The formulation of foods with controlled sensory properties remains, therefore, a challenge. The influence of texturing agents has not been well elucidated. Their main effect is a modification of the viscosity, sometimes resulting in a significant decrease in perceived flavour [1]. The given explanation is that increased viscosity hinders the mass transfer of flavour compounds to the surface of the food product [2]. However, thickened solutions of similar viscosity do not induce the same flavour perception. Furthermore, some studies showed that although thickening of solutions affected flavour intensities, it did not result in a change in the in-nose measured flavour concentration [3,4]. Moreover, one study claims that the texture of gels determines perception of volatile flavour intensity rather than in-nose flavour concentrations [4]. An interaction between the composition and structure/texture of the food on one hand and oral processing on the other hand is likely to play an important role in these phenomena. In the present study, the influence of oral processing protocol on *in vivo* flavour release was examined for two strawberry flavoured custard desserts which differed in texture. The texture was varied by adjusting the concentration of CMC in the custards. A group of 21 volunteers was used, and the in-nose measurements of the strawberry flavour compounds were carried out by PTR-MS.

2. Materials and Methods

Chemicals. A commercial strawberry flavour mixture was obtained from Givaudan (Duebendorf, Switzerland). Its composition was published previously [5]. Ethyl butyrate, ethyl iso-pentanoate and ethyl hexanoate were present at the following concentrations 90 mg/g, 10 mg/g and 20 mg/g, respectively. High viscosity CMC (C-5013; Sigma-Aldrich Chemie, Steinheim, Germany) was used for custard preparation.

Custard preparation. Two different custards were prepared. They were composed of 0.1% and 1.0% CMC, respectively. For custard preparation, 936 (0.1% CMC custard) or 927 (1.0% CMC custard) full-fat milk was heated to 60°C in a water bath. Sucrose (63 g; Siucra; Irish Sugar Ltd, Carlow, Ireland) was added and the mixture stirred for 3 min. The CMC was added in small increments to ensure that the CMC was fully dispersed. To obtain the custard texture, the mixture was stirred again for 5 min. The temperature of the water bath was increased to 95°C, while stirring continued. When the custard reached a temperature of 90°C, heating continued for another 10 min. The custard was subsequently cooled down at room temperature for 15 minutes and further down to 30°C by placing the bottle in cold water. Forty g of the custard was placed in a 100 ml glass bottle, 14 µl of the flavour mixture was injected in the custard and the bottle sealed. Final total flavour concentration was 56 mg/kg custard. The mixture was stirred for 5 min and stored at 6°C for 24 h prior to analysis. For each type of custard duplicate batches were prepared.

In-nose analysis. For in-nose analysis, a fork-shaped glass nosepiece was placed with its two inlets in the nostrils of each subject. The nosepiece had one outlet for breathing and an orthogonal outlet for sampling. The latter was used to remove the air, without disturbing the assessor's breathing or eating pattern. The air was drawn in at a rate of 100 ml min^{-1}, 15 ml of which was led into the PTR-MS. The background was measured for 30 s. During that time an assistant placed seven g of custard (20°C) on a spoon. The assessor transferred the custard to his/her mouth. Two different oral processing protocols were applied. Subjects were either allowed to chew and swallow freely, or they had to follow instructions (10 movements protocol). The instructions involved moving the custard ten times left to right to left in their mouths during 15 s. They immediately swallowed after the 15 s. Preliminary scans (mass range m/z 30-220) of the flavoured custards as well as the individual flavour compounds revealed that the masses m/z 117, m/z 131 and m/z 145 could be exclusively assigned to ethyl butyrate, ethyl iso-pentanoate and ethyl hexanoate, respectively. The other compounds were either below detection limits or had parent/major product ions in common. Twenty-one subjects participated in the in-nose analyses. Two batches of the individual custards were analysed (2 replicates per type of custard per chewing protocol per person). The samples were analysed according to the method described by Lindinger and co-workers [6]), while employing a constant drift voltage of 600 V. Transmission of the ions through the quadrupole was considered according to the specification of the instrument. The spectra were background and transmission corrected. From the individual curves, maximum intensities (Imax), time to maximum intensities (t(Imax)), and total release (area) were determined. During the in-nose experiments, the time to swallowing (t(swallow)) was determined as well.

Statistical analysis. The Imax, t(Imax), area and t(swallowing) data were subjected to multivariate analysis of variance (MANOVA) to determine significant differences between the custards, the chewing protocols, and the assessors. Principal component analysis (PCA) was conducted on the Imax data set. A significance level of $P<0.05$ was used throughout the study.

3. Results and Discussion

In-nose analysis was carried out on two custards varying in texture, using two oral processing protocols. From the real time curves, Imax, t(Imax) and area values were calculated. They are presented in Table 1 together with the t(swallow) data.

Overall, the type of compound affected all parameters significantly. Highest Imax and area values were obtained for ethyl butyrate, followed by ethyl iso-pentanoate and ethyl hexanoate. t(Imax) followed the same order, for both oral processing protocols. Ethyl hexanoate was found in lower concentrations in the expired breath of subjects than ethyl iso-pentanoate, whereas the concentration of ethyl hexanoate was twice as high as the concentration ethyl iso-pentanoate in the custard. It was shown before that this is due to the higher affinity of this larger, more hydrophobic compound for the matrix [5]. The presence of milk fat may play a role.

MANOVA also showed that the CMC concentration in the custard had a significant overall effect on t(Imax) and t(swallow), but not on Imax and the area. When looking at the data for the

Table 1
Three strawberry flavour compounds released from custards varying in texture (0.1 and 1.0% CMC) using two oral processing protocols (free and 10 movs): results from in-nose PTR-MS analysis (Imax, t(Imax), area, t(swallow); mean ± SD)[*]

	Ethyl butyrate		Ethyl iso-pentanoate		Ethyl hexanoate	
	0.1% CMC	1.0% CMC	0.1% CMC	1.0% CMC	0.1% CMC	1.0% CMC
Imax						
10 movs	270±193	284±209	24±17	25±17	8.0±4.9	8.0±4.7
free	256±190	226±156	21±15	18±11	7.5±4.2	6.7±6.7
t(Imax)						
10 movs	30±7	31±5	32±7	33±7	40±8	42±9
free	19±3	21±5	21±4	23±5	28±5	33±12
Area						
10 movs	6106±4405	5808±3495	625±412	654±356	302±189	292±179
free	4312±2872	4302±2497	452±275	434±231	238±137	210±190
t(swallow)						
10 movs	14±4	15±5	14±4	15±5	14±4	15±5
free	4±2	6±3	4±2	6±3	4±2	6±3

[*]CMC is carboxymethyl cellulose; 'free' is free chewing and swallowing protocol; '10 movs' is the imposed regime protocol; Imax is maximum intensity; t(Imax) is time to maximum intensity, area is total amount of flavour measured; t(swallow) is time to swallowing.

individual compounds and samples (Table 1-2), it is shown that Imax increased when the CMC concentration was increased using the 10 oral movements protocol. When the subjects were allowed to chew freely, the 0.1% CMC custard showed highest Imax values. These differences between the custards with different textures were, however, not significant (Table 2) due to the high variance in the measurements. Increase in CMC concentration led to higher t(Imax) values (Table 1), although this was only significant for ethyl hexanoate. No significant effect of CMC concentration on the area values was observed. The t(swallow) values were significantly increased by increase in CMC concentration in the custards for both oral processing protocols. The higher concentration CMC resulted in higher variance for both oral processing protocols. This indicates that differences between individuals become more pronounced for samples with a more complex texture, even when an imposed oral processing regime is used.

The oral processing protocol influenced t(Imax), area and t(swallow) significantly. Generally, free chewing resulted in lower Imax, t(Imax), area and t(swallow) values than the imposed regime protocol. These effects were all significant except for the ethyl butyrate and ethyl hexanoate Imax values. During natural eating, the custards were swallowed fairly quickly (<4 s). The imposed regime protocol involved mouth movements during 15 s. This was significantly longer than the time to swallowing when subjects consumed the custards as they normally would. It is not surprising that this prolonged oral processing resulted in increased intensities and increased persistence of the strawberry flavour. The imposed regime protocol resulted in lower variance in the measurements. The coefficients of variance for the free chewing and imposed regime protocol were 52 and 32%, respectively.

Table 2
Analysis of variance results: probability levels [%] associated with F-values of the three factors CMC concentration, chewing protocol and subjects for Imax, t(Imax), area and t(swallow) data of in-nose PTR-MS analysis of three strawberry flavour compounds released from custards*

	CMC concentration	Chewing protocol	Subjects
Imax			
Ethyl butyrate	70.1	9.1	**0.0**
Ethyl iso-pentanoate	52.9	**1.7**	**0.0**
Ethyl hexanoate	64.9	33.5	**0.7**
t(Imax)			
Ethyl butyrate	22.0	**0.0**	**2.2**
Ethyl iso-pentanoate	11.9	**0.0**	7.3
Ethyl hexanoate	**1.1**	**0.0**	5.6
Area			
Ethyl butyrate	62.9	**0.0**	**0.0**
Ethyl iso-pentanoate	87.5	**0.0**	**0.0**
Ethyl hexanoate	44.0	**0.4**	**0.0**
t(swallow)			
All compounds	**0.0**	**0.0**	**0.0**

*Codes are explained in Table 1. In bold: significant probabilities at a 5% level.
The subjects showed significant differences for all parameters. For the individual compounds, significant differences between subjects were found for all compounds for Imax, area and t(swallow) values, as well as for t(Imax) of ethyl butyrate. PCA showed that the subjects could be divided into two groups: some released consistently more strawberry flavour from the thicker custard (high CMC concentration), and some from the more liquid custard (low CMC concentration). More research is needed to understand the underlying conditions.

4. Conclusions
This study showed significant effects of oral processing protocols on in-nose concentrations of strawberry flavour compounds during eating of custard desserts. It also showed interaction between oral processing and custard texture and its effects on in-nose strawberry flavour concentrations.

5. References
1 Clark. In Frontiers in Carbohydrate Research; R. Chandrasekaran, ed.; Elsevier Applied Science: New York, NY, 1992; pp 85.
2 E. R. Morris. In Food Polysaccharides and their Application; A.M. Stephen, ed.; M. Dekker: New York, NY, 1995; pp. 517.
3 T.A. Hollowood, R.S.T. Linforth, A.J. Taylor. Chem Senses 27 (2002) 583.
4 K.G.C. Weel, A.E.M. Boelrijk, A.C. Alting, P.J.J.M. van Mil, J.J. Burger, H. Gruppen, A.G.J. Voragen, C. Smit. J Agric Food Chem 50 (2002) 5149.
5 S.M. van Ruth, L. de Witte, A. Rey. J Agric Food Chem (in press)
6 W. Lindinger, A. Hansel, A. Jordan. J Mass Spectrom Ion Proc 173 (1998) 191.

VOCs Monitoring by PTR-MS as a Method to Assess Agronomical Procedures:
The Test Case of Golden Delicious Ripening

M. Vescovi[1], A. Weber[1], D. Barbon[1], A. Tonini[1], L. Fadanelli[2], A. Dorigoni[2], G. Stoppa[3], R. Verucchi[1], S. Iannotta[1] and A. Boschetti[1]

[1]Institute of Photonics and Nanotechnology, ITC Division, Via Sommarive 18, 38050 Povo di Trento – Italy – abosche@itc.it

[2]Istituto Agrario di S. Michele all'Adige, 30010 S. Michele all'Adige, Trento Italy.

[3]Dipartimento di Informatica e Studi Aziendali, Università degli Studi di Trento, Via Inama 5- 38100 Trento Italy

ABSTRACT

We report on studies devoted to verify the potential of PTR-MS monitoring of VOC as a method to assess the effects of agronomic procedures in the cultivation of golden delicious apples. The test cases proposed are: the evaluation of the effects on preservation and shelf life of the harvesting time (before, during and after the optimal time) and of chemically controlled ripening by two different plant growth regulators. The comparison with laser photoacoustic detection of ethylene, the gas plant hormone, shows that PTR-MS can indicate markers for such processes.

1. Introduction

The long term preservation of apples is strongly depending on the initial state of the product. In particular the degree of ripening affects the final quality and its depletion during preservation and the other marketing phases including shelf life. The optimization of harvesting time would require that most of the fruit should be gathered in a very short time which is not easily done in an economic way. A complementary approach would be to optimize the marketing procedures for the different degrees of ripening of the harvested fruits. In this framework an increasingly relevant role is being given to plant growth regulators that could control the timing of the fruit ripening.

To this end it is very important to have information about the way the different harvesting times and chemicals affect the fruits and their quality. Our approach is to use VOCs emissions monitored by PTR-MS (1,2,3) and Photoacoustic (PA) laser spectroscopy (4,5,6) to address these questions.

2. Materials and Methods

For the experiments dealing with the harvesting time, the fruit samples, golden delicious apples coming from the Bassa Val di Non, a valley of Trentino-Italy specifically devoted to the production of these fruits, have been selected in 2002 by the IASMA in collaboration with the producers (APOT). Three different harvesting times were selected corresponding to early (27/09), optimal (04/10) and late (11/10) ripening with the respect to standards as determined by the studies of IASMA. After harvesting, the apples were preserved under standard controlled atmosphere (1.5% O2, 2.5% CO2) and temperature (1.2 °C). The samples were than taken out from the storage rooms at three different periods of preservation that is 1.5, 4 and 5.5 months. The VOC were measured on the headspace of 4 single apples for each period

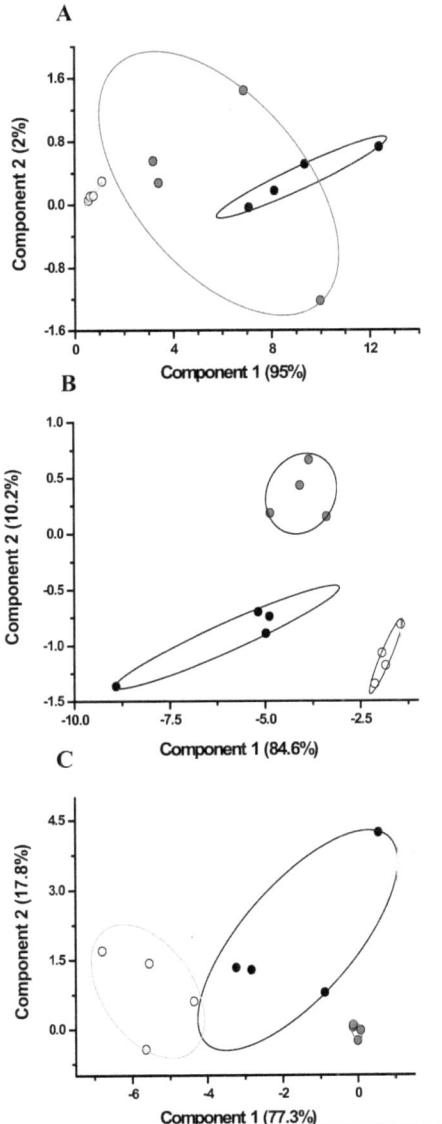

A

B

C

Figure 1 Relative PCA on selected masses. The evolution over 15 days shelf-life after 1.5 months of preservation. A) first day, B) 7th day, C) 15th day. The different data sets refer to the harvest time: early (○), optimal (●), late (●). The ellipses are draw at 95% of confidence level.

and harvesting time. In order to assess the shelf life evolution at room temperature the same samples were monitored the day of extraction from the storage rooms and after 7 and 15 days.

The treatment with the plant growth regulator (PRG) Ammino Etoxy Vinil Glicine (AVG), used to retard the ripening process, was carried out in an experimental field run by IASMA. Three different stage of growth of the apples, that is 40, 30, 20 days before the programmed harvest time, were considered for studying the effect of the treatment.

The other treatment, considered for reducing the natural loss of crop load on the trees during ripening, has been carried out in the same field. The Naphthyl Acetic Acid (NAA) was the Auxina compound used.

Four apples were selected and harvested from very similar trees being treated in the same way every 5-7 days along the whole ripening process between 15/08 and 15/10. For comparison 4 apples were also harvested at the same times from non-treated plants having very similar characteristics. Such procedure was optimized in order to avoid any side effect such as modifications of tree crop load.

3. Results and Discussion

Figure 1 shows the shelf life evolution of apples harvested at the three different stages of ripening (early, optimal and late) and then preserved for 1.5 months. The R-PCA analysis (7), based on a set of masses selected by the variance, shows a clear discrimination of the different samples. We observed a similar ability of discrimination over all the other preservation times studied.

Figure 2 shows the evolution of the emission for three different volatile organic compounds during the long period of the ripening process running from the middle of August up to the middle of October. In particular ethylene, the gas hormone of plants, is a very interesting one that cannot be easily monitored by PTR-MS. We therefore studied it by PA laser spectroscopy. The effects of the different NAA and AVG treatments mentioned above, are reported. Ethylene is recognized marker of the ripening process and shows clearly the onset of ripening occurring

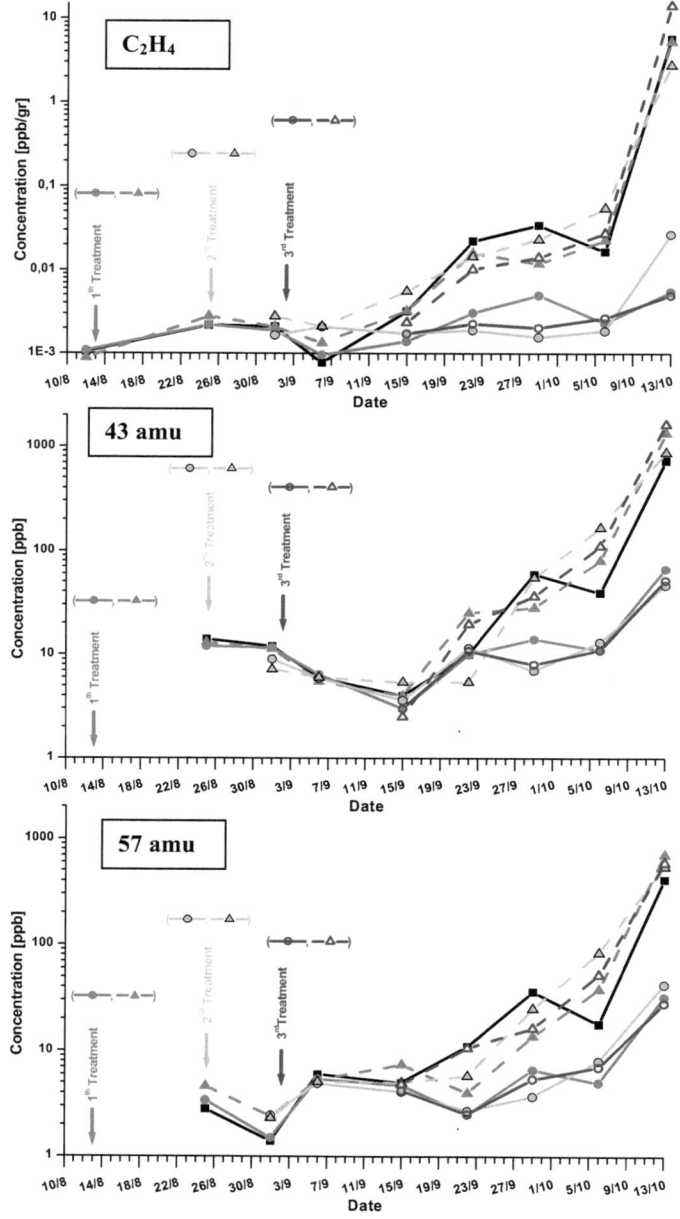

around the middle of September. It also discriminate that the NAA treated samples have a trend very similar to the untreated one. On the other hand the AVG treated apples have a much delayed evolution of the emission corresponding to a slower ripening process.

The PTR-MS measurement demonstrated the presence of masses such as 43, 57, 71, 85, 89 amu that show very similar trends. This is strong indication that PTR-MS could be used to monitor effects of ripening induced by chemical process. A better understanding of the underlying mechanisms requires further studies

Figure 2 Evolution the emission of three VOC during the ripening process of apples. Solid lines with data as circles refer to the AVG treated samples while dashed lines with data as triangles refer to the NAA treated samples. The different gray tones indicate the time at which the treatment was carried our as reported in the figure. The solid line, with the data in black squares, reports the results for samples of untreated apples.

4. Acknowledgement

This work was funded by the Autonomous Province of Trento, Fondo Unico 2001, Project Qualiquant (Relationships between fruit crop load and quality in apple: an interdisci-plinary approach), Project Mugo (MUltiGas detection by Opto-acoustic-buildup cavity laser spectroscopy) and by the "Convenzione PAT-CNR" Project: Analysis and researches on the agro-industrial system (Subproject3).

5. References

1. Lindinger W., Hansel A. and Jordan A,." Proton-transfer-reaction mass spectrometry (PTR-MS) : on line monitoring of volatile organic compounds at pptv levels." Chem. Soc. Rev **27** (1998), 347-354.

2. Boschetti A., Biasioli F., van Opbergen M., Warneke C., Jordan A., Holzinger R., Prazeller P., Karl T., Hansel A., Lindinger W. and Iannotta S. "PTR-MS real time monitoring of the emission of volatile organic compounds during post-harvest ageing of berry fruit." Postharv Biol Technol **17** (1999), 143-151.

3. Biasioli, F., Boschetti, A., Toccoli, T., Jordan, A., Fadanelli, L., Lindinger, W. and Iannotta, S. "Proton transfer reaction mass spectrometry : a new technique to assess post harvest quality of strawberries." Proc. 4th Int. Strawberry Symp. Acta Hort. **567** (2002), 739-742.

4. Harren, F.J.M., Reuss, J., Woltering, E.J. and Bicanic, D.D.. "Photoacoustic measurements of agriculturally interesting gases and detection of C_2H_4 below the ppb level." Appl. Spectros. **44** (1990),1360-1367.

5. Harren, F.J.M. and Reuss, J. " Spectroscopy, photoacoustic. Encyclopedia of Appl. Phys. **19** (1997), 413-427.

6. Boschetti A., Bassi D., Jacob E, Iannotta S., Ricci R., Sotoni M. "Resonant-photoacoustic simultaneous detection of methane and ethylene by means of a NIR diode laser," Appl. Phys. B **74** (2002), 273-278.

7. Barbon D., Weber A., Vescovi M., Tonini A., Boschetti A., Iannotta S., Fadanelli L. and Stoppa G.,. "A statistical approach for the analysis of proton transfer reaction mass spectrometry (PTR-MS) data aimed a a qualification of fruits based on VOC emissions." Proc. 5th Int. Postharvest Symp. Verona, Italy 6-11 June (2004).

Linking PTR-MS spectra with molecular markers: a new tool to control fruit quality in marker assisted selection

Elena Zini[1], Franco Biasioli[1], Flavia Gasperi[1], Daniela Mott[1], Eugenio Aprea[1,2], Tilmann D. Märk[2], Andrea Patocchi[3], Cesare Gessler[3] and Matteo Komjanc[1]

[1] *Istituto Agrario San Michele all'Adige, Via E. Mach 2- 38010 San Michele all'Adige (TN)-Italy (Elena.zini@iasma.it)*

[2] *Institut für Ionenphysik, Universität Innsbruck, Technikerstrasse 25, 6020 Innsbruck-Austria*

[3] *Institute of Plant Science, ETH, Universitätstrasse 2- 8092 Zürich-Switzerland*

Volatile Organic Compounds (VOCs) play an important role in fundamental and applicative issues related to fruit and, more generally, to agroindustrial products both because they are directly related to their perceived sensory quality and because they allow a non invasive and fast monitoring of many metabolic and physiological processes induced, e.g., by ripening, storage and stress conditions like wounding, disease or plant-insect interaction
The availability of genetic linkage maps enables the detection and analysis of QTLs (Quantitative Trait Loci) contributing to quality traits of plant genotype. This is the basis for Marker Assisted Selection (MAS): that is the use of markers flanking a gene of interest to allow the selection for the presence (or absence) of a gene in a new progeny and to follow the inheritance of genes, particularly those which cannot be readily identified.
In the case of volatile compounds the main limitations to the use of MAS are the lack of information regarding the identification of markers related to VOCs and the fact that the effect of volatile compounds in, e.g., sensory quality is not easily modeled and predictable.
A possible way to address this problem is to find an instrumental characterization of fruit crop that is linkable both to the desired quality attribute and to genetic information.
We showed elsewhere that Proton Transfer Reaction Mass Spectrometry (PTR-MS) can efficiently be used to classify strawberry genotypes by non destructive measurements and that PTR-MS spectra are correlated with sensory characterization of agroindustrial products. Based on these indications we decided to investigate if it is possible to find molecular markers related to the PTR-MS fingerprint of fruits and thus provide a possible link between fruit quality, sensory quality in particular, and molecular characterization.
PTR-MS has been used to analyze the headspace composition of the VOCs emitted by apple genotypes of the progeny 'Fiesta' X 'Discovery'. Samples have been characterized by their PTR-MS spectra normalized to total area. QTL analysis for all PTR-MS peaks has been performed on the linkage map of a subset of 'Fiesta' x 'Discovery' population (57 genotypes, two fruits per genotype). 10 genomic regions associated with the peaks at m/z=28, 43, 57, 61, 103, 115 and 145 have been detected (LOD>2.5).
In conclusion we show that it is possible to find QTLs related to PTR-MS fingerprint of the headspace composition of single whole apple fruits indicating the presence of a link between molecular characterization and PTR-MS data. Chemical information is available and we provide tentative identification of the metabolites related to the detected QTLs. A relation between apple skin color and peaks related to carbonylic compounds was identified.

AKNOWLEDGMENTS

We would like to acknowledge the support of the MIUR-MURST project QUALIFRAPE (Italy) for partly financing DM and the FWF (Wien). We would like to thank Robert Liebhard for his encouraging help and for his useful indications and Mauro Jertmini for providing the samples

PTR-MS measurement of headspace concentration of volatile organic compounds after wounding and during eating of strawberry fruits

Franco Biasioli[1], Flavia Gasperi[1], Eugenio Aprea[1,2], Daniela Mott[1], Tilmann D. Märk[2]

[1] *Istituto Agrario San Michele all'Adige, Via E. Mach 2- 38010 San Michele all'Adige (TN)-Italy (Elena.zini@iasma.it)*

[2] *Institut für Ionenphysik, Universität Innsbruck, Technikerstrasse 25, 6020 Innsbruck-Austria*

Odor and aroma perception is a complicated process that involves many human receptors and their response to the stimuli induced by their contact with the mixture of volatile compounds that reach the receptors of the retro-nasal cavity. To better understand this process is of outmost interest to have a direct real time monitoring of the genuine VOCs concentration that our receptors are exposed to. In their review (Flavor release, ACS Symposium Series, 763, 2000) on the techniques for measuring volatile release in vivo during consumption of food A. Taylor and R. Linforth indicate three main possibilities to address this issue: Atmospheric Pressure Ionization Mass Spectrometry (API-MS), Proton Transfer Reaction Mass Spectrometry (PTR-MS) and selective photo-ionization.

Odor and flavor are relevant features that are recognized by the consumer and that have several industrial applications. This is particularly true for strawberries.

In this work we used PTR-MS to preliminarily investigate the release of volatile compounds during eating and to study the role of wound compounds on the perceived odor/flavor of strawberry.

The effects related to cultivar (Miss and Queen Elisa) and shelf life (12-120 hours) are considered. We show that, beside a rapid increase of most PTR-MS peaks after cutting, the response to mechanical stress (cut) of PTR-MS signals can be classified in two groups. The first one comprises spectrometric peaks showing slow time evolution (relatively stable in time for hours). Their behavior (intensity and time evolution) does not depend on the presence of air or pure nitrogen (endogenous compounds). Some of them are related to shelf life and not to cultivar (e.g.:m/z=59), Other are strongly dependent on cultivar E.g.: m/z=103, mostly methyl butanoate, is always 1-2 orders of magnitude higher in Miss than in Queen Elisa. The second group of spectrometric peaks shows a relatively fast increase (several minutes), probably related to oxidative stress, leading to the production of many compounds that can tentatively be associated with the hexenal and hexanal family (e.g., m/z=57, 81, 99). For these latter peaks the signal is strongly reduced when the experiment is performed under nitrogen (oxidation products).

Nose space measurements indicate that these compounds seem to be relevant also in the composition of the air exhaled during strawberry consumption and thus potentially important for their perceived flavor. We have evidence that the observed relative intensities in these in vivo experiments are quite different from those measured in model systems. Data are compatible with literature data obtained by APCI.

Our work indicates, on the one hand, the possibility of monitoring these compounds by PTR-MS in real time during induced mechanical stress or during eating and, on the other hand, that a proper study and comprehension of the correlation between sensory perception and instrumental determination of volatile concentration can not be based only on model systems but needs to be carried out also in vivo.

Work supported by the Italian MIUR-MURST (QualiFraPe project), the Austrian FWF of Vienna, and the European Commission.

PTR-MS and microarray analysis of strawberry: an integrated metabolomic and functional genomic approach

Fabrizio Carbone[1], **Fabienne Mourgues**[1], **Filomena Giorno**[1], **Franco Biasioli**[2], **Flavia Gasperi**[2], **Tilmann D. Maerk**[3], **Gaetano Perrotta**[1], **Carlo Rosati**[1]

[1] *ENEA Trisaia - S.S. 106, km 419+500 - 75026 Rotondella (MT) – Italy (carlo.rosati@trisaia.enea.it)*

[2] *Istituto Agrario San Michele all'Adige, Via E. Mach 2- 38010 San Michele all'Adige (TN)-Italy*

[3] *Institut für Ionenphysik, Universität Innsbruck, Technikerstrasse 25, 6020 Innsbruck-Austria*

Anonymous PTR-MS (Proton Transfer Reaction-Mass Spectrometry) fingerprinting of fruit samples (strawberry and apples) provides a fast and non-destructive tool for cultivar classification and for comparison with other anonymous characterization means as molecular markers.

Here, on the contrary, we investigate the possibility to exploit the quantitative chemical information present in the PTR-MS spectra as a basis for a metabolomic study of strawberry and for its correlation with functional genomic approaches.

In particular we studied the transcriptome of the octoploid strawberry (Fragaria x ananassa) in a number of Italian élite varieties (here only data regarding the cultivar Queen Elisa and her parents Miss and USB359 are shown) to assess fruit quality and correlate gene expression data with those from biochemical analyses and panel tests. More than 3000 expressed sequence tags (ESTs) obtained from a fruit cDNA library were analyzed. The 50 largest contigs related to plant genes with assigned functions comprised 725 ESTs (23.7% of total), including genes not only associated with ripening (cell wall metabolism, sugar and acid synthesis, pigment formation, vitamin synthesis and allergenic properties), but also with plant-pathogen interactions, abiotic stresses and housekeeping functions. Over 1800 selected ESTs were used to produce a cDNA microarray, together with ripening-related candidate genes and checks. Comparative profiling experiments revealed a limited number of genes with variations in expression levels among different genotypes. Nevertheless, such differentially expressed genes were related to important quality and post-harvest traits as fruit firmness and aroma profile, and data from microarray experiments were confirmed by results of Real Time PCR analyses. PTR-MS analysis of the head space composition of whole fruits of studied genotypes suggested a correlation between the presence in the fruit of some classes of metabolites and the expression of genes putatively involved in their synthesis. PTR-MS classification of fruits showed to be robust to year-to-year variation and culture technique. New results from ongoing molecular and biochemical analyses and correlation with data from sensory analysis will point out the genes and compounds most correlated to fruit quality and post-harvest traits, as well as overall consumers' preferences.

This preliminary integrated functional genomics and metabolomics approach complements other studies where PTR-MS data have been related to genetic information and indicates that this fast technique can play an important role in the investigation of aroma biosynthesis and its genetic basis.

ACKNOWLEDGMENTS

Work supported by the Italian MIUR-MURST (QualiFraPe project), the Austrian FWF of Vienna, and the European Commission. We thank ISF-FO staff for data on quality traits, A. Falzone for the web utilities of EST collection, and M.A. Carboni for her assistance.

Advanced oxidation in olive oil: monitoring of secondary reaction products and detection of rancid defect

Eugenio Aprea[1,2], **Franco Biasioli**[1], **Flavia Gasperi**[1], **Graziano Sani**[3], **Claudio Cantini**[3], **Tilmann D. Märk**[2]

[1]*Istituto Agrario di S. Michele a/A, S. Michele, Via E. Mach, 2, 38010, Italy (eugenio.aprea@iasma.it)*

[2]*Institut für Ionenphysik, Universität Innsbruck, Technikerstr. 25, A-6020 Innsbruck, Austria*

[3]*Istituto per la Valorizzazione del Legno e delle Specie Arboree, Consiglio Nazionale delle Ricerche, via Madonna del Piano, Sesto Fiorentino, 50019 Italy*

ABSTRACT

In the present contribute we propose the direct measure of volatile organic compounds (VOCs) in olive oil headspace by means of Proton Transfer Reaction Mass-Spectrometry (PTR-MS) aiming at the on line monitoring of the oxidation state and at the detection of rancid defect. A thermoxidation process has been applied to extravirgin olive oils samples and, contemporarily to headspace monitoring, peroxide value has been measured. A model for the prediction of peroxide value from PTR-MS signal has been built and verified by cross-validation. The PTR-MS technique provided a fast, non invasive and sensitive method to follow the oxidation process. Moreover, defective (rancid) olive oils, characterised by a trained panel, were also measured and data compared with thermoxidised samples. Multivariate data analysis of PTR-MS spectra allows the non-ambiguous classification of defective and extravirgin oils.

1. Introduction

Lipolysis and oxidation are the processes leading to the most serious deterioration of olive oil. The first is a natural process that starts when the oil is still in the fruit and increase with ripening for the action of endogenous enzymes, while oxidation begins after the oil is extracted from the fruit. Lipids oxidation is the result of deterioration of the fatty acids, in particular the unsaturated ones, when catalytic agents (light, enzymes, metallic cations, etc.) are present. The first steps of lipids oxidation led to the formation of hydroperoxides, that are odourless, colourless, tasteless and very instable compounds. From the hydroperoxides decomposition (advanced oxidation) secondary reaction products as aldehydes, ketones, acids, alcohols and hydrocarbons are formed. Some of these volatile compounds often adversely affect flavour, odour, taste, nutritional value, and overall quality of oil [1]. From a sensory point of view this produces a rancid defect that negatively affect the pleasant characteristic of extravirgin olive oil.

The most used methods to determine the oxidation state of oils are related to the measurements of the concentration of primary and/or secondary oxidation products. Among those based on the concentration of primary oxidation products, peroxide value, which measures hydroperoxide concentration, is one of the most widely used. Other methods, are based on the concentration of secondary oxidation products including aldehydes, ketones, acids, alcohols, lactones, ethers, hydrocarbons and furan derivatives. One example is the measure of nonanal and hexanal in the oil headspace [2].

Here we use PTR-MS to monitor volatile formation in oil during induced thermoxidation. Data were calibrated on peroxide measures and the volatile fraction of the oils was measured before, during and after the thermoxidation. Oil samples with rancid defect, characterised by a trained sensory panel, were also measured and multivariate analysis was used both to classify samples (extravirgin/defective) and to compare them with thermo-oxidized samples.

2. Experimental

Bottles (120 ml, Supelco, Bellefonte, USA) containing 2.5 ml of oil, extravirgin olive oil (19 samples in duplicate) and defective (rancid) olive oil (10 samples in duplicate) as classified by a trained sensory panel, were closed and kept at 28°C for one hour before and during the measurements. The VOCs released in the headspace were transferred trough a heated (70 °C) capillary line, realized in uncoated deactivated fused silica tubing with inner diameter of 0.25 mm (Supelco, Bellefonte, USA), directly into the drift tube of the PTR-MS (PTR-MS FDT, Ionicon Analytik Ges.m.b.H., Innsbruck, Austria) at a rate of 10 sccm exchanging it with pure nitrogen gas (SOL, Verona Italy; purity: 99.999%). Four samples for each of the extravirgin olive oils, closed in bottles (120 ml, Supelco, Bellefonte, USA), were placed in a convention oven at the temperature of 110 °C and measured over a time span of about 30 hours. Peroxide value of oils was measured with a FOODLAB portable analyser (CDR, Ginestra Fiorentina, Firenze, Italy, http://foodlab.cdr-mediared.it).

3. Results and discussion

The intensities of several peaks are highly correlated with peroxide value. Excluding their isotopes, we found 10 masses with a correlation coefficient r>0.9 (m/z: 125, 143, 111, 103, 61, 127, 73, 101, 97, 139). Comparing the pattern fragmentation of pure compounds, partly reported by Buhr et al. [3] and partly measured in our laboratories, with data on volatiles reported in the literature for olive oil we can give a tentative attribution of spectrometric peaks. The two masses with the highest correlation with peroxide value (m/z 125 and 143) are compatible with nonanal (the correlation between the two masses is 0.93). Nonanal concentration increases in virgin olive oil during oxidation time as showed by Morales et al. [2]. It is a secondary oxidation product that derives from the homolytic β-scission of the 9-hydroperoxide and 10-hydroperoxide oleate, two primary oxidation products.
In the panel a) of fig. 1 is shown how the signal at m/z 125 increases with thermoxidation proceeding and in the second panel (b)) the signal at m/z 125 is plotted against peroxide value. Similar behavior is observed for m/z=143 (data not shown). Fragment with m/z 111 is a possible marker for octanal originated from β-scission of the 11-hydroperoxide oleate and 1-octen-3-ol, a product of further oxidation. Fragments at m/z 97 and 101 are possible markers respectively for heptanal originated from 11-hydroperoxide linoleate and hexanal originated from 13-hydroperoxide linoleate. Signal at m/z 139 could be a measure of nonadienals (i.e. 3,6-nonadienal from 9-hydroperoxide linolenate), while signal at m/z 127 could be a measure of 2-octenal originate from 11-hydroperoxide linoleate. Signal at m/z 61 could be attributed to acetic acid. All these attributions, even if tentative, are in good agreement with literature indication and compatible with chemical processes involved in lipids oxidation.
Peroxide value of thermoxidised oils, was calibrated on entire PTR-MS spectra collected (from m/z 20 to m/z 260). The best model was developed using 13 pls (fig.2) showing the minimum root square means error in prediction. The model accounts for 94.75% of the explained variance in validation.
In fig. 3 the plot of first two components of the discriminant Partial Least Square analysis is reported, showing a good separation between defective and extravirgin olive oil samples. A supervised technique like discriminant Partial Least Square [4] seems more appropriate for this kind of analysis because the difference extravirgin/defective is not the major source of

145

variability. Cross-validation have been used to check the ability of the model to properly classify unknown samples: one or more samples have been removed from the data set and a model, built with the remaining samples using the first and second dPLS scores, was used to classify them. A 100% correct classification using linear discriminant analysis (Mahalanobis distance) on only 2 dPLS scores was obtained. We have also indication that the multivariate approach adopted even in previous studies [5,6] gives good performances in the discrimination of cultivar in monovarietal oils (data not shown).

4. Conclusions

It is possible to follow the oxidative processes induced by heating through headspace monitoring by means of PTR-MS. The intensity of many masses increases during this process and many of them are strongly correlated with the independent determination of the peroxide value. The most correlated peaks are those probably associated with aldehydes that are the main secondary product of oil oxidation [2]. The coupling of PTR-MS spectra with chemometric techniques (multivariate calibration in this case) allows the accurate determination of peroxide value and oxidation state of oils based only on PTR-MS data providing a fast and non invasive tool for the detection and the measurement of oils rancidity. Indication of the possibility of discriminate defective oils and monovarietal oils is suggested.

5. Acknowledgements

Work partially supported by the PAT-CNR project AGRIIND and by the FWF, Wien, Austria.

6. References

1. Vercellotti, J. R.; St. Angelo, A. J.; Spanier, A. M. (1992). Lipid Oxidation: an Overview. In Lipid Oxidation in Foods; St. Angelo, A. J., Ed.; ACS Symposium Series 500; American Chemical Society: Washington, DC; pp 1-11.
2. Morales, M. T., Rios, J. J., & Aparicio, R. (1997). Changes in the volatile composition of virgin olive oil during oxidation: flavours and off-flavors. Journal of Agriculture and Food Chemistry, 45, 2666-2673.
3. Buhr, K., van Ruth, S., & Delahunty, C. (2002). Analysis of volatile flavour compounds by Proton Transfer Reaction-Mass Spectrometry: fragmentation patterns and discrimination between isobaric and isomeric compounds. International Journal of Mass Spectrometry 221 (1), 1-7.
4. Kemsley, E.K. (1998). Discriminant Analysis and Class Modelling of Spectroscopic Data; John Wiley & Sons Ltd: Chichester,UK.
5. Biasioli F., Gasperi F., Aprea E., L. Colato, Boscaini E., Märk T. (2003). Fingerprinting mass spectrometry by PTR-MS: heat treatment vs. pressure treatments of red orange juice – a case study. Int. J. Mass Spectrom., 223-224, 343-353.
6. Biasioli F., Gasperi F., Aprea E., Mott D., Boscaini E., Mayr D., Märk T.D. (2003). Coupling Proton Transfer Reaction-Mass Spectrometry with Linear Discriminant Analysis: a Case Study. J. Agric. Food Chem., 51, 7227-7233.

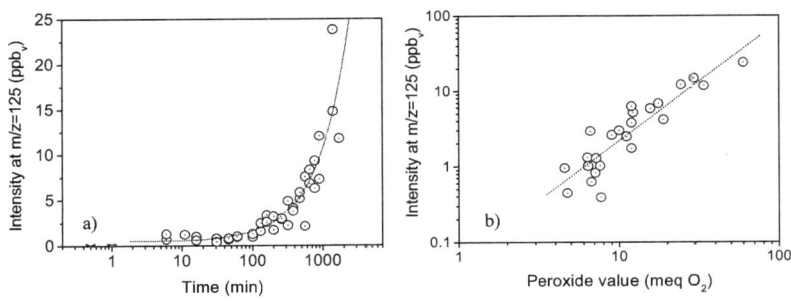

Fig. 1. a) Temporal evolution of signal at m/z 125 during olive oil thermoxidation process. **b)** Olive oil peroxide value plotted vs intensity of the PTR-MS signal at m/z 125.

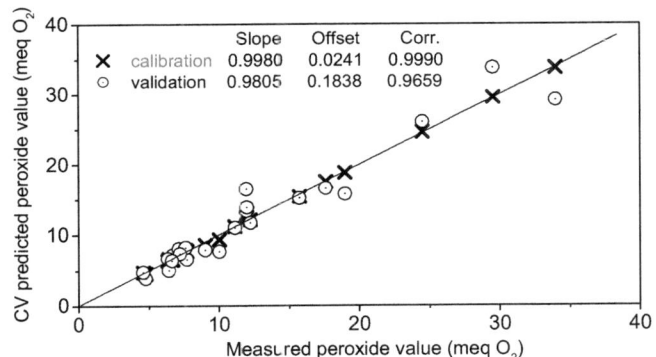

Fig. 2. Peroxide value estimation based on PTR-MS data.

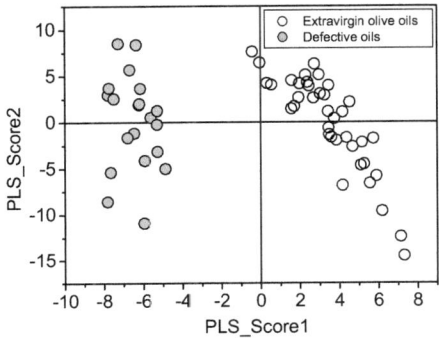

Fig. 3. Discriminant Partial Least Square analysis of PTR-MS spectra of defective and extravirgin olive oils.

Estimation of overall flavor and odor sensory profile of hard cheese from PTR-MS data

F. Gasperi[1], F. Biasioli[1], E. Aprea[1,2], D. Mott[1], I. Endrizzi[1], V. Framondino[1], S. Carlin[1], G. Versini[1], F. Marini[3], T. D. Märk[2]

[1]*Istituto Agrario San Michele all'Adige, Via E. Mach 2- 38010 San Michele all'Adige (TN)-Italy (Flavia.gasperi@iasma.it)*

[2] *Institut für Ionenphysik, Universität Innsbruck, Technikerstrasse 25, 6020 Innsbruck-Austria*

[3] *Dipartimento di Chimica, Università "La Sapienza", P.le A. Moro 5, 00185 Roma, Italy*

Descriptive analysis by a panel of trained judges is probably the best way to objectively assess and compare the sensory properties of food products. It is a widespread and important technique because it is the most direct method to investigate the perceived quality of food and, coupled with consumer studies, plays a crucial role in the understanding of their preference and of its drivers. In particular for typical products, i.e. products with a designation of origin or related to traditional manufacturing processes, consumer appreciation in relation with sensory profiles must be studied to exploit and protect their high added value. Among the most used and consolidated descriptive techniques providing qualitative and quantitative information we follow here the conventional profiling by Quantitative Descriptive Analysis (QDA). Being QDA expensive and time consuming, food scientists and technologists have constantly tried to find correlation between QDA and instrumental data to reduce the necessary use of QDA only to a preliminary calibration phase. Proton Transfer Reaction-Mass Spectrometry (PTR-MS) has been used in several fields, from environmental to medical applications, and it is getting more and more recognized as an important tool for detection of volatile organic compounds. Recently its application in relation with sensory analysis has been proposed. In particular in our laboratories we applied PTR-MS to different food products indicating the possibility of a fast and accurate quality control and finding interesting correlation with sensory analysis.

Here we present a complete procedure that provides the possibility to estimate the overall flavor/odor sensory profile of 20 samples of "Trentingrana", the variety of Grana Padano produced in Trentino (Northern Italy) on the basis of direct PTR-MS spectral characterization. A panel of 8 trained judges performed the sensory analysis following a vocabulary of 30 attributes. Only attributes related to odors (6 attributes) and flavors (6 attributes) have been considered here. At the same time analytical information on the chemical compounds responsible for the observed features is directly provided by the loading associated to the latent variables produced by the calibration models. Details on the data analysis of PTR-MS spectra and their correlation with sensory data are described.

In particular results of descriptive statistics are shown and the performances of different multivariate calibration methods (Multi-linear regression, Partial Least Squares, both PLS1 and PLS2) are compared by evaluating the errors in the cross-validated estimation of the sensory attributes. PLS2 seems to give a good average description providing an overall insight of the problem but does not perform well in the prediction of the individual sensory attributes. PLS1 analysis is more accurate and has reasonable errors in most cases but it uses several latent variables making the interpretation of the loadings not straightforward. The preliminary application of Orthogonal Signal Correction (OSC) filtering on PTR-MS spectra followed by

PLS1 analysis results in a good estimation for most of the attributes and has the advantage to use only 1 or 2 latent variables. Results of SPME-GC/MS analysis on similar samples are compatible with the model developed on the basis of PTR-MS data and with attribution of spectrometric peak to chemical compounds.

Because we have evidence that the spectral PTR-MS fingerprint can be directly related to the effects induced by agroindustrial processes and to genetic features of fruits we suggest that it can be a useful bridge between the sensory evaluation of food and other characterizations tools providing a fast and non destructive method to support product development, process control, breeding, etc.

Acknowledgements: Work partially supported by PAT-CNR project AGRIIND, by PAT project RASO and by the FWF, Wien.

Characterization of wine with PTR-MS

Elena Boscaini, Tomas Mikoviny, Armin Wisthaler, Eugen von Hartungen, Nooshin Araghipour, and Tilmann D. Märk

Institut für Ionenphysik, Universität Innsbruck, Technikerstrasse 25, A-6020 Innsbruck, Austria; tilmann.maerk@uibk.ac.at

A new method for measuring volatile profiles of alcoholic beverages (or other ethanol-containing analytes such as perfumes or herbs) has been developed. The method is based on proton-transfer-reaction mass spectrometry (PTR-MS). However, instead of hydronium ions (H_3O^+) protonated ethanol clusters ($C_2H_5OH_2^+(C_2H_5OH)_{n=1,2}$) are used as chemical ionization reagent ions. A stable reagent ion distribution is obtained by a 10-fold dilution of analyte headspace into ethanol saturated nitrogen. Samples with different ethanol content can thus be directly compared. Characteristic mass spectral fingerprints have been obtained for four wine varieties. Principal component analysis discriminates between different wine varieties and shows specific correlations between wine variety and selected ions.

Microbially mediated emissions of VOCs from organic waste

T. Mikoviny[1], A. Wagner[2], S. Mayrhofer[2], A. Wisthaler[1], H. Insam[2], A. Hansel[1], T.D. Märk[1]

[1]*Institut für Ionenphysik, Technikerstr 25, A-6020 Innsbruck, Austria (tomas.mikoviny@uibk.ac.at)*

[2]*Institut für Mikrobiologie, Technikerstr. 25d, A-6020 Innsbruck, Austria*

Collection of source-separated organic household wastes is standard procedure in several European countries. However, malodour emissions that are mediated by microbial activity during early-phase degradation of the various substrates are problematic. This study investigated the relationship between the composition of the microbial community in different types of organic waste bins with the rate and patterns of volatile organic compound (VOC) emissions measured by Proton-Transfer-Reaction Mass Spectrometry (PTR-MS). Standardised mixtures of organic wastes were stored in 3 different types of 120 L waste bins: a standard waste bin; a standard waste bin with a biodegradable bag to prevent odour release; and a standard waste bin with a biodegradable bag and a biofilter lid to reduce odour emissions and inhibit spore germination. Bacterial and fungal communities were extracted at two-day intervals and microbiologically characterized. In an attempt to identify specific VOC producers, PTR-MS was used additionally to measure emissions from pure cultures of fungi, bacteria and yeasts growing on standard media for 27 hours. First results of the ongoing research will be presented.

PTR-MS measurements of carboxylic acids: axillary odour investigations and determination of Henry's law constants

Eugen von Hartungen[1], Armin Wisthaler[1], Dagmar Jaksch[1], Elena Boscaini[1], Patrick J. Dunphy[2], Julia Märk[1] and Tilmann D. Märk[1]

[1] *Institut für Ionenphysik, Universität Innsbruck, Technikerstrasse 25, A-6020 Innsbruck, Austria; eugen.hartungen@uibk.ac.at*

[2] *Danisco (UK) Limited, Wellingborough, Northants, England*

Proton-Transfer-Reaction Mass Spectrometry (PTR-MS) was used as an analytical tool to measure gas-phase concentrations of short-chain fatty acids. Chemical ionisation of C_2-C_6 carboxylic acids by PTR-MS produced intense protonated molecular ions (with traces of hydrates) along with acylium ion fragments. Gas-phase concentrations were derived using the established method for calculating PTR-MS sensitivity factors. Henry's law constants of carboxylic acids for aqueous solutions at 40°C were determined. Direct monitoring of volatile fatty acids, known to be associated with secretions from the human axilla, was performed via a specially designed transfer device situated in the axilla of the test subjects. Mass spectral data corresponded with the findings of a sensory assessor.

Atmospheric chemistry of C_3-C_6 cycloalkanecarbaldehydes

Barbara D'Anna[1], Armin Wisthaler[2], Øivind Andreasen[1], Jyrki Viidanoja[3], Niels R. Jensen[3], Claus J. Nielsen[1], Armin Hansel[2], Jens Hjorth[3], Yngve Stenstrøm[4]

[1] Department of Chemistry, University of Oslo, Oslo, Norway

[2] Institut für Ionenphysik, Leopold-Franzens-Universität Innsbruck, Innsbruck, Austria (armin.wisthaler@uibk.ac.at)

[3] European Commission, DG – Joint Research Centre, Institute for Environment and Sustainability, Climate Change Unit, Italy

[4] Department of Chemistry, Biotechnology and Food Science, Chemistry Section, Agricultural University of Norway, Aas, Norway

The rate coefficients for the gas phase reaction of NO_3 and OH radicals with a series of cycloalkanecarbaldehydes have been measured in purified air at 298 ± 2 K and 760 ± 10 Torr by the relative rate method using a static reactor equipped with long-path FT-IR detection. The values obtained for the OH radical reactions (in units of 10^{-11} cm^3 molecules^{-1}s^{-1}) were: cyclopropanecarbaldehyde, 2.13 ± 0.05; cyclobutanecarbaldehyde, 2.66 ± 0.05; cyclopentanecarbaldehyde, 3.27 ± 0.07; cyclohexanecarbaldehyde, 3.75 ± 0.05. The values obtained for the NO_3 radical reactions, (in units of 10^{-14} cm^3 molecules^{-1}s^{-1}) were: cyclopropanecarbaldehyde, 0.61 ± 0.04; cyclobutanecarbaldehyde, 1.99 ± 0.06; cyclopentanecarbaldehyde 2.55 ± 0.10; cyclohexanecarbaldehyde, 3.19 ± 0.12. Furthermore, the reaction products with OH were investigated using long-path FT-IR spectroscopy and proton-transfer-reaction mass spectrometry (PTR-MS). The identified products cover a wide spectrum of compounds including nitroperoxycarbonyl cycloalkanes, cycloketones, cycloalkyl nitrates, multifunctional compounds containing carbonyl, hydroxy and nitrooxy functional groups, HCOOH, HCHO, CO and CO_2.

Disjunct Eddy Covariance Measurements of Monoterpene Fluxes From a Norway Spruce Forest using PTR-MS

W. Grabmer[1], M. Graus[1], C. Lindinger[1], A. Wisthaler[1], B. Rappenglück[2], R. Steinbrecher[2], A. Hansel[1]

[1]Institut für Ionenphysik, Universität Innsbruck, Technikerstraße 25, 6020 Innsbruck, Austria
wolfgang.grabmer@uibk.ac.at

[2]Forschungszentrum Karlsruhe GmbH, Institut für Meteorologie und Klimaforschung, Atmosphärische Umweltforschung (IMK-IFU), Kreuzeckbahnstraße 19, 82467 Garmisch-Partenkirchen, Germany

Interest in reliable quantification of organic trace compounds released from terrestrial ecosystems stems from their impact on oxidant levels such as ozone and hydroxyl radicals, and on secondary organic aerosol formation. In an attempt to quantify these emissions, a disjunct sampler (DS) was coupled to a PTR-MS instrument as shown in figure 1. The system was fixed on top of a tower about 10 m above the canopy of a Norway spruce forest (Waldstein, Germany), during the second intensive field campaign of BEWA2000 in summer 2002.

The disjunct eddy covariance (DEC) method is a derivative of the eddy covariance (EC) technique. EC requires the assumption of a constant flux layer within the lowest part of the planetary boundary layer, which is normally formed at turbulent conditions during daytime. Instead of the continuous sampling of EC, DEC utilises an instantaneous grab sample taken at intervals of tens of seconds, with vertical wind speed recorded at the instant of sample collection (Lenschow et al., 1994; Rinne et al., 2001; Warneke et al. 2002). Intermittent periods are used for sample analysis by a moderately fast chemical sensor, in this case a PTR-MS instrument. The vertical turbulent transport of a trace compound is then calculated from the covariance of the fluctuations in compound volume mixing ratio (VMR) and vertical wind speed, which reflect turbulent eddies of different sizes. Fluxes of monoterpenes below 0.5 nmol m^{-2} s^{-1} were detected (Fig. 2). Results were in the same order as data obtained using relaxed eddy accumulation and the enclosure approach (Graus, 2004; Grabmer et al., 2005), and were in good agreement with reported emission rates of Norway spruce in literature (Kesselmeier and Staudt, 1999).

In addition to the field experiment, laboratory tests were carried out to validate the disjunct sampling procedure: The same DS was fixed to a Teflon bag, which was flushed with variable concentrations of VOC from a gas standard and monitored with a second PTR-MS instrument (Fig. 3). As can be seen in fig. 4, no measurable carry-over or memory effects were observed for monoterpenes or even for oxygenated species such as methanol.

This work has been published in the International Journal of Mass Spectrometry (Grabmer et al., 2004)

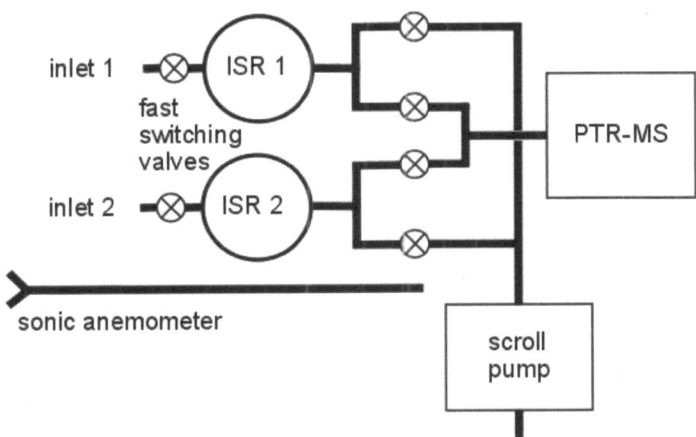

Fig. 1. Schematic diagram of the DS unit

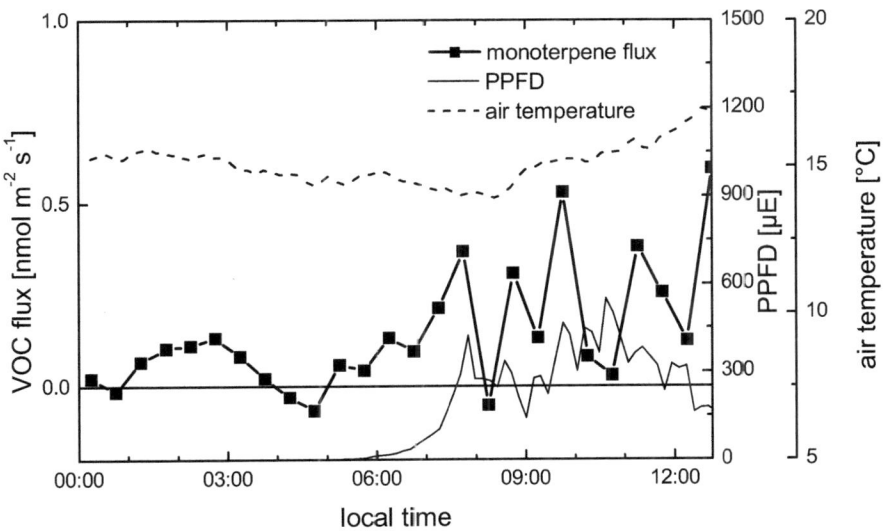

Fig. 2. Half-hour means of monoterpene flux rates as measured on 10th August, 2002. PPFD is photosynthetic photon flux density.

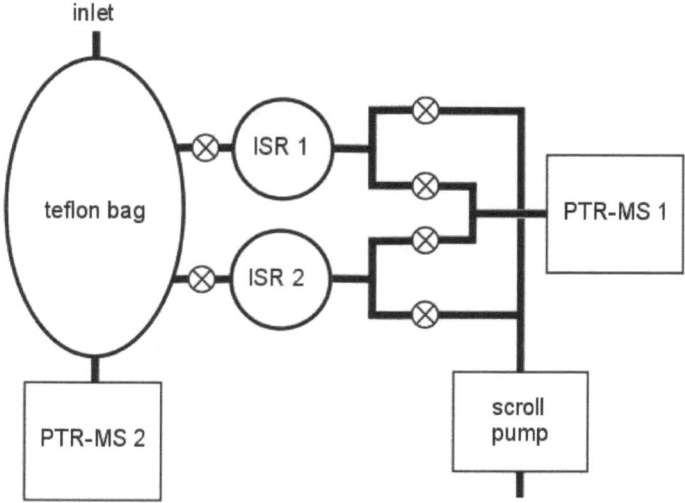

Fig. 3. Set-up of the laboratory experiment. Variable VOC VMRs were simulated in a Teflon bag and monitored by a second PTR-MS system. The rest of the set-up is identical to that of the field experiment (Fig. 1)

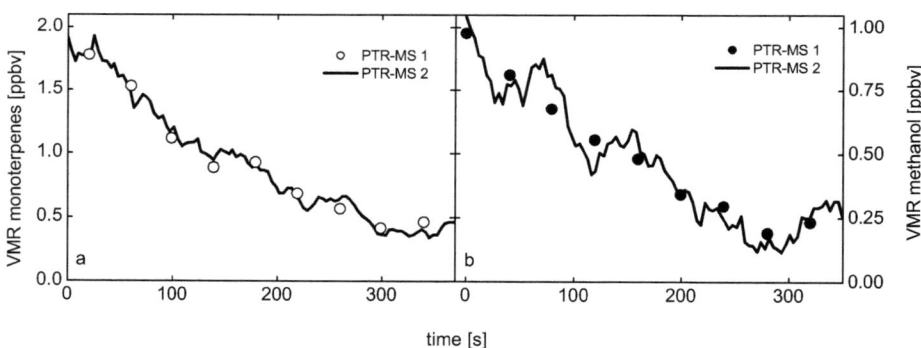

Fig. 4. Results of the laboratory experiment for monoterpenes (left) and methanol (right). Circles indicate VOC VMRs alternately taken in ISR1 and ISR2 and determined by PTR-MS 1. Solid lines represent VOC VMRs in the Teflon bag as measured continuously by PTR-MS 2.

References

Grabmer W., J. Kreuzwieser, C. Cojocariu, M. Graus, H. Rennenberg, D. Steigner, R. Steinbrecher, A. Wisthaler, A. Hansel, 2005. VOC Emissions from Norway Spruce (Picea abies L. [Karst]) Twigs in the Field - Results of a Dynamic Enclosure Study, submitted to Atmos. Environ.

Grabmer W., M. Graus, C. Lindinger, A. Wisthaler, B. Rappenglück, R. Steinbrecher, A. Hansel, 2004. Disjunct Eddy Covariance Measurements of Monoterpene Fluxes From a Norway Spruce Forest using PTR-MS, Int. J. Mass Spectrometry, in print

Graus M., 2004, personal communication

Kesselmeier J., Staudt M., 1999. Biogenic Volatile Organic Compounds (VOC): An Overview on Emission, Physiology and Ecology, Journal of Atmospheric Chemistry 33, 23-88.

Lenschow D. H., Mann J., Kristensen L., 1994. How Long Is Long Enough When Measuring Fluxes and Other Turbulence Statistics? Journal of Atmospheric and Oceanic Technology 11, 661-673.

Rinne H. J. I., Guenther A. B., Warneke C., de Gouw J. A., Luxembourg S. L., 2001. Disjunct eddy covariance technique for trace gas flux measurements, Geophysical Reseach Letters 28, 3139-3142.

Warneke C., Luxembourg S. L., de Gouw J. A., Rinne H. J. I., Guenther A. B., Fall R., 2002. Disjunct eddy covariance measurements of oxygenated volatile organic compounds fluxes from an alfalfa field before and after cutting, J. Geophys. Res.107, 4067-76.

PTR-MS as a Technique for Investigating Stress Induced Emission of Biogenic VOCs

J. Beauchamp[1], A. Hansel[1], E. Kleist[2], M. Miebach[2], U. Weller[2], A. Wisthaler[1], J. Wildt[1]

[1]*Institut fuer Ionenphysik, Leopold-Franzens-Universitaet, A-6020 Innsbruck, Austria*
Jonathan.Beauchamp@uibk.ac.at

[2] *Institut Phytosphere (ICG-III), Forschungszentrum Juelich, D-52425 Juelich, Germany*

ABSTRACT

Proton-transfer-reaction mass spectrometry (PTR-MS) was used in conjunction with two GC-MS systems to investigate stress induced emissions of volatile organic compounds (VOCs) from plants. Experiments were performed in the laboratory under well defined conditions and VOC emissions were induced by ozone exposure at variable concentrations, for different durations, and with different light conditions. Tobacco (*Nicotiana tabaccum* var. Bel W3) plants were used as the investigated species due to their known sensitivity to ozone stress.
VOCs measured included methanol, C6- alcohols and aldehydes, methyl salicylate and sesquiterpenes. Results show quantitative linear temporal responses of plants to ozone stress that are dependent upon the amount of stress encountered. Furthermore, it was determined that the flux density of ozone taken up by the plant, rather than ozone concentration or AOT40 values, was a much more accurate reference for the plants' responses to ozone stress.
This investigation demonstrated further the ability of PTR-MS to provide excellent high time-resolved measurements of the relevant species, allowing determination of such temporal responses of plants to stress.
Measurement technique and experimental results will be presented.

1. Introduction

Emissions of volatile organic compounds (VOCs) by vegetation are estimated to be about an order of magnitude higher than those of anthropogenic origin (e.g. Fehsenfeld et al., 1992; Mueller, 1992; Guenther et al., 1993). VOCs have a strong impact on chemical processes in the atmosphere such as ozone production or formation of aerosols (Kavouras et al., 1998; Guenther et al., 2000) and thus, plant-generated VOC emissions play a central role in atmospheric chemistry.
Biogenic VOC emissions commonly depend on light and temperature (e.g. Guenther et al., 2000; Niinemets et al., 2004 for review), but additionally respond to stress conditions (e.g. leaf wounding, water and temperature stresses, and pathogen attack). To investigate possible relationships between VOC emissions and amount of stress, ozone exposure was used as elicitor for VOC emissions. The use of ozone as a stress agent is advantageous for a number of reasons: exposures can be conducted under well-defined conditions; experiments can be repeated using the same amount of ozone, thus applying the same amount of stress to plants; the amount of stress caused by ozone exposure can be varied over a wide range, allowing investigation of the plants' responses in relation to the degree of stress.
The aim of this investigation was therefore to test a hypothesis that VOC emissions from tobacco plants are quantitatively related to the degree of ozone exposure.

2. Experimental

Tobacco plants (*Nicotiana tabaccum*, var. Bel W3) were used for these experiments due to their particular sensitivity to ozone. 9 to 10 week old plants were individually placed in a continuously stirred tank reactor (as described by Wildt et al., 1997). The plants were exposed to ozone (produced via oxygen photolysis under a UV lamp) in short pulses (1 h – 8 h) and with a range of concentrations (80 to 1700 ppbv). Additionally, various light regimes were implemented. High time resolution measurements using proton-transfer-reaction mass spectrometry (PTR-MS; Lindinger et al., 1998) allowed the dynamics of the plants' responses to ozone stress to be studied, with GC-MS aiding compound identification.

3. Results

Net photosynthesis and transpiration rates were observed to diminish during exposure to ozone, indicating the closure of the plants' stomata (fig. 1). Consequently, ozone flux into the plant also decreased.

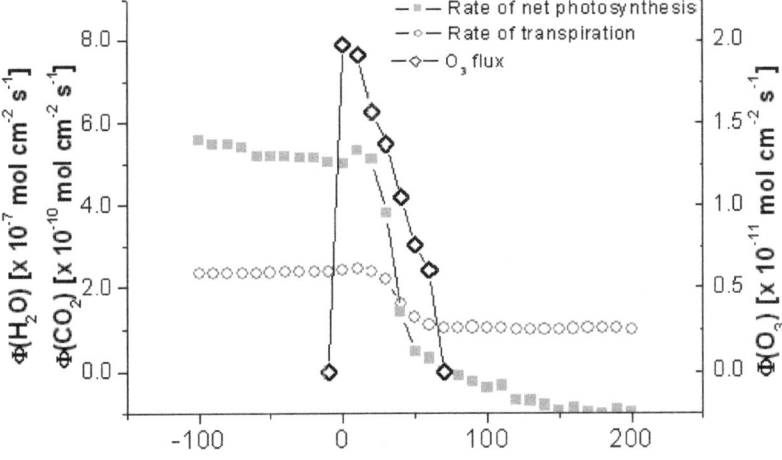

Figure 1. Rates of net photosynthesis and transpiration, as well as ozone flux into the plant, during periods of ozone exposure.

After ozone exposure, a series of VOCs were emitted by the plants. These generally followed a similar pattern in all cases of exposure (fig. 2): Methanol (MeOH, at m/z 33) emissions came first and were most abundant, describing two large peaks; sesquiterpene (SQT, at m/z 205) and methyl salicylate (MeSa, at m/z 153) emissions followed thereafter, typically with one or two peaks and in lower abundances; the products of lipoxygenase activity (sum of LOX products, from the octadecanoid pathway, at m/z 83+85+99+101) came last, but in great abundance.

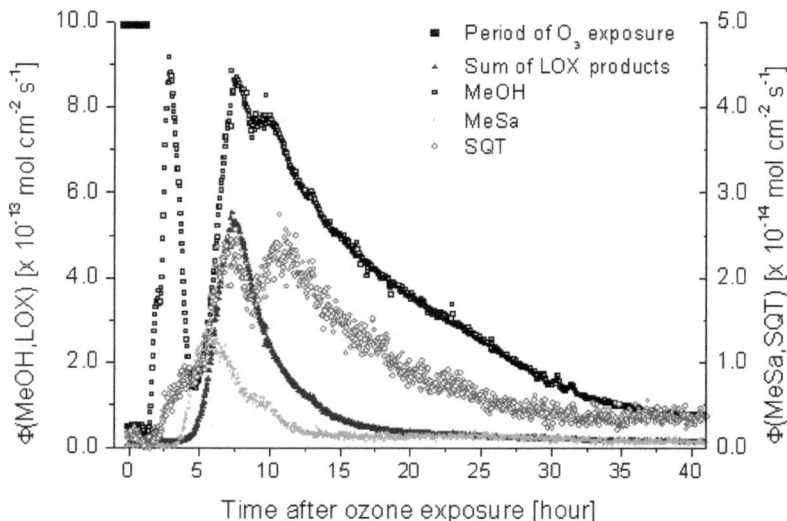

Figure 2. Typical time evolution of VOC emissions from tobacco Bel W3 after exposure to ozone (ozone exposure period indicated in upper left-hand corner of plot).

A focus of this study was the VOC emissions associated with lipoxygenase (LOX) activity within the plants. The high time resolved measurements of these emissions allowed investigation of the initial plant response to ozone exposure. It was found that the pattern of increase of these compounds followed a sigmoidal behaviour, with the inflexion point of this shape being designated lag time, D, and its inverse being defined as induction rate. Data of induction rates were plotted against ozone flux density into the plant (fig. 3). Results indicate a linear relationship between induction rate and ozone flux density, with a regression coefficient of $R^2 = 0.74$. An absolute response of the plants was also found, with five ozone molecule taken up leading to approximately one LOX-product VOC emitted.

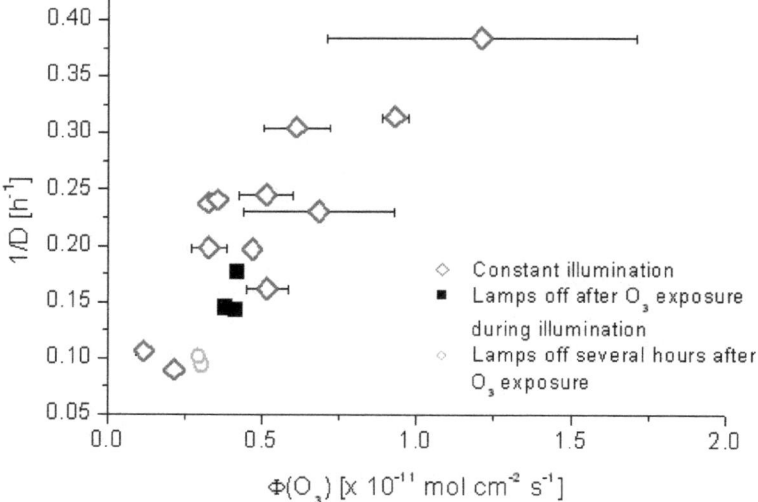

Figure 3. Induction rate (1/D, where D is the lag time from the start of ozone exposure to inflexion point of LOX product emissions) versus ozone flux density taken up by plant.

4. Conclusion

This study revealed that there is a quantitative link between stress-induced VOC emissions from plants and the amount of stress applied. Furthermore, results identified an absolute response of tobacco (Bel W3 variety) to the ozone flux density into the plant. This leads to the conclusion that the amount of ozone flux is the better reference for plant stress response, than is total ozone concentration or AOT40 values.

This investigation further demonstrates the applicability of combined PTR-MS and GC-MS measurements to allow for detailed investigation of the dynamics of plants' responses to ozone stress.

A detailed report of this work has been submitted to the Journal "Plant, Cell and Environment".

References

Fehsenfeld, F., Calvert, J., Fall, R., Goldan, P., Guenther, A.B., Hewitt, N., Lamb, B., Liu, S., Trainer, M., Westberg, H. and Zimmerman, P. (1992) Emissions of volatile organic compounds from vegetation and the implications for atmospheric chemistry. *Global Biogeochemical Cycles* 6, 389-430

Guenther, A., B., Zimmerman, P.R., Harley, P.C., Monson, R.K. and Fall, R. (1993) Isoprene and monoterpene emission rate variability: model evaluation and sensitivity analysis. *Journal of Geophysical Research* 98, 12609-12617

Guenther, A., Geron, C., Pierce, T., Lamb, B., Harley, P., Fall, R. (2000) Natural emissions of non-methane volatile organic compounds, carbon monoxide, and oxides of nitrogen from North America. *Atmospheric Environment* 34, 2205-2230

Kavouras, I.G., Mihalopoulos, N., Stephanou, E.G. (1998) Formation of atmospheric particles from organic acids produced by forests. *Nature* 395, 683-686

Lindinger, W., A. Hansel, and A. Jordan, Proton-transfer-reaction mass spectrometry (PTR-MS): on- line monitoring of volatile organic compounds at pptv levels, *Chemical Society Reviews* 27, 347-354, 1998

Mueller, J.-F. (1992) Geographical distribution and seasonal variation of surface emissions and deposition velocities of atmospheric trace gases. *J Geophysical Research* 97, 3787-3804

Niinemets, Ü., Loreto, F., Reichstein, M. (2004) Physiological and physico-chemical controls on foliar volatile organic compound emissions. *Trends in Plant Science* (In Press)

Wildt, J., Kley, D., Rockel, A., Rockel, P., Segschneider, H.J. (1997) Emission of NO from several higher plant species. *Journal of Geophysical Research* 102, 5919-5927

Development of a fully automated PTR-MS for measurements onboard a passenger aircraft (CARIBIC-Project)

Detlev Sprung and Andreas Zahn

Institute of Meteorology and Climate Research, Forschungszentrum Karlsruhe, Hermann-von-Helmholtz-Platz 1, 76344 Eggenstein-Leopoldshafen, Germany, detlev.sprung@imk.fzk.de

ABSTRACT

Within the project CARIBIC II (Civil Aircraft for Regular Investigation of the atmosphere Based on an Instrument Container) a Proton-Transfer-Reaction Mass Spectrometer system was modified for automated and continuous on-line measurements of volatile organic compounds (VOC) onboard a passenger aircraft. The concept of CARIBIC is to use regularly operating passenger aircraft for measurements of atmospheric trace constituents at cruising altitude of 8-12 km, i.e. in the upper troposphere and lower stratosphere (UTLS) [Brenninkmeijer, et al., 1999]. This region belongs to the least well-understood areas of the atmosphere, mainly because of its extreme dynamical complexity and the scarcity of accurate data. In comparison to research aircraft used during short-term campaigns this concept is a relative cost-effective alternative for collection of detailed information about the UTLS over a time horizon of more than 10 years.

Within CARIBIC II altogether ~60 trace gases and ~20 aerosol parameters are detected using an Airbus 340-600 operated by Lufthansa AG as off December 2004 on a monthly basis. The PTR-MS will be operated for measurements of acetone, acetaldehyde, methanol, and acetonitrile. Important modifications of the PTR-MS include completely new software, inter alia to allow the integration into an Ethernet network in which all 16 instruments installed in the CARIBIC container can communicate with each other. Also a full automation was realized, e.g. to control the pumps and the quadrupole mass spectrometer or to monitor pressures and temperatures at the most important locations in the system. Due to safety instruction for passenger aircrafts extensive mechanical and electrical changes had to be performed. To detect the expected trace gas concentrations with sufficient accuracy and precision, the system performance was improved, e.g. by integrating a 3rd turbo molecular pump, a new high voltage supply, the heating control of important parts and an online calibration system. The system performance will be presented on the poster.

1. Introduction

Within CARIBIC a reinforced airfreight container equipped with different scientific instruments is installed into the cargo bay of a passenger aircraft. During CARIBIC I between November 1997 and April 2002 about 75 flights were successfully performed onboard a

Boeing 767 operated by LTU International Airlines [Zahn et al., 2002]. For CARIBIC II the air carrier changed. Now Lufthansa provides an Airbus 340-600 with an aircraft endurance of 14300 km. Measurements over long distances during return flight are intended. A contract with Lufthansa will allow long-term measurements for more than 10 years. Compared to CARIBIC I a bigger airfreight container was installed providing the possibility to operate more scientific equipment.

Figure 1: The instrument container in the configuration installed on the Airbus 340-600 (total weight ~1.5 t) for CARIBIC II (BPS: Basic power supply, TRU: transfer rectifier unit, DOAS: Differential optical absorption spectrometer, OPC: Optical particle counter, CPC: channel particle counter, PTRMS: proton transfer reaction mass spectrometer)

The container built up for CARIBIC I carried instruments for measuring carbon monoxide (Max Planck Institute for chemistry, MPIC, Mainz, Germany), ozone (Institute meteorology and climate research, IMK, Forschungszentrum Karlsruhe, FZK, Germany), aerosol concentration (Institute for tropospheric research, IfT, Leipzig, Germany) and nitrogen oxides (NO, NO_y) (German Aerospace Center, DLR, Oberpfaffenhofen, Germany). Furthermore aerosol elemental composition was measured (University of Lund, Sweden) and canister samples were taken, which were subsequently analyzed in the laboratory for ~15 hydrocarbons (mainly alkanes and alkenes), the isotopic composition of CO and CO_2 (MPIC), and halocarbons (University of East Anglia, UEA, UK). Detailed meteorological support such as back-trajectory trajectory analyses was provided by the Royal Netherlands

Meteorological Institute (KNMI, de Bilt, Netherlands). The bigger CARIBIC II container (Figure 1) contains additional measurements, inter alia, for measuring CO_2 (Centre de la Recherche Scientific, CNRS, Gif-Sur-Yvette, France), BrO, CH_2O, NO_2, SO_2 (University Heidelberg, Germany), H_2O (for water vapor and cloud water, IMK), Hg (GKSS research center, Geesthacht, Germany), O_2/N_2 (University of Bern, Switzerland) and VOCs by using the PTR-MS described here (IMK). Of prime interest are acetone, acetonitrile, methanol and acetaldehyde. The scientific interest on these compounds, especially in the UTLS, was described elsewhere [e.g. Bange and Williams, 2000; Singh et al., 1995, 2001; Wennberg et al., 1998].

Exchange between troposphere and stratosphere, long range transport and convective transport will be particularly investigated during CARIBIC [Zahn et al., 2000]. This project will give the chance to investigate processes on global and seasonal time scales overcoming the problems of spot measurements during short-term campaigns which only provide an instantaneous state of the atmosphere. The long measurement period of ~10 years allows the comparison of the present and future composition of the UTLS. A distinction between natural and anthropogenic sources of diverse chemicals is expected. The data are achieved for comparisons of models and satellite data. With the help of atmospheric chemistry transport models, the CARIBIC data will help to better quantify (i) the budgets of many trace constituents in the UTLS, (ii) emission rates of the US, Europe, Africa, and Asia (by comparing measurements upwind and downwind of these continents), (iii) the influences of the Indian and Mexican monsoon on the (sub-) tropical UT, (iv) the influence of biomass burning, e.g. the deforestation in Africa, South America and Asia, (v) tracer fluxes between troposphere and stratosphere, and (vi) the influence of air traffic on certain trace gas budgets, e.g. NO_x and O_3, of the UTLS.

2. Modifications on the PTR-MS system

To meet the strict safety regulations required for the use of instruments onboard passenger aircraft, to work with limited power consumption and to be able to detect the low VOC concentrations present in the UTLS, extensive modifications of the commercially available PTR-MS of IONICON [Hansel et al., 1995; Lindinger et al., 1998] had to be performed.

The required strict safety regulations motivated many of these modifications. A power management was necessary, i.e. the power control of many individual system components, partially because the power supply of the container was divided into two parts, a basic power supply (BPS) providing 24 VDC and a transfer rectifier unit (TRU) providing 28VDC. Only at cruise (above 500 m above ground) the TRU is switched on. EMC regulations had to be fulfilled. An air proved rack was used; sensitive parts were put on shock mounts to protect them against vibrations and shocks. Flight phases and power management are controlled by a master PC that communicates via an Ethernet system with the slave computers of the different instruments. For allowing to include this communication into the operational software of the PTR-MS as well as the control of certain subsystems (pumps, T- and p-regulators), the monitoring of new system parameters (pressures, temperatures, fluxes) sensors and the operation of the mass spectrometer itself, a completely new operational software based on LapView was written. In case of a PC failure an electronic emergency off was integrated (not controlled by the software).

To meet the demands on the expected low trace gas concentrations, accuracy and precision of the data the PTR-MS was upgraded with a third turbo pump for enhancing the working pressure in the drift tube and the efficiency of the ion-molecule reactions leading to the trace gas detection. The enhancement of the drift tube pressure requires an adjustment of the electrical field. Therefore a new high voltage supply for the lenses in the drift tube was built up to provide an optimal E/N ratio (electrical field E, Ion concentration N). An online calibration system was also introduced, providing a calibration gas standard during the flights as well as a low power Pt-catalyst to determine the background signal of the measured mass spectrum. A new inlet system was created which consists of several valves, flow and pressure controllers, capillaries and heaters to provide controlled and constant conditions during the flight, independent on ambient pressure and the actual container temperature.

References:

Bange H. and J. Williams, New directions: Acetonitrile in atmospheric and biogeochemical cycles, Atmos. Environ., 34, 4959-4960, 2000.

Brenninkmeijer, C. et al., CARIBIC – Civil aircraft for global measurement of trace gases and aerosols in the tropopause region, J. Atmos. Oceanic. Technol., 16, 1373-1383, 1999.

Hansel, A., A. Jordan, R. Holzinger, P. Prazeller, W. Vogel, and W. Lindinger, Proton-transfer-   reaction mass spectrometry (PTR-MS): on-line monitoring of volatile organic compounds   at volume mixing ratios of a few pptv, Int. J. Mass Spectrom. Ion Processes, 149/150,   609-619, 1995.

Lindinger, W., A. Hansel, and A. Jordan, On-line monitoring of volatile organic compounds at   ppt-levels by means of proton-transfer-reaction mass spectrometry (PTR-MS): Medical   applications, food control and environmental research, Int. J. Mass Spectrom. Ion   Processes, 173, 191-241, 1998.

Singh, H.B., M. Kanakidou, P.J. Crutzen, and D.J. Jacob, High concentrations and photochemical fate of oxygenated hydrocarbons in the global troposphere, Nature, 378,   50-54, 1995.

Singh, H., Y. Chen, A. Staudt, D. Jacob, D. Blake, B. Heikes, and J. Snow, Evidence from the   Pacific troposphere for large global sources of oxygenated organic compounds, Nature,   410, 1078-1081.

Wennberg, P.O., et al., Hydrogen radicals, nitrogen radicals, and the production of ozone in the   upper troposphere, Science, 279, 49-53, 1998.

Zahn, A. et al., Identification of extratropical two-way troposphere-stratosphere mixing based   on CARIBIC measurements of O_3, CO, and ultrafine particles, J. Geophys. Res., 105,   1527-1535, 2000.

Zahn, A., C.A.M. Brenninkmeijer, P.J. Crutzen, G. Heinrich, H. Fischer, J.W.M. Cuijpers, and   P.F.J. van Velthoven, Distributions and relationships of O_3 and CO in the upper   troposphere: The CARIBIC aircraft results 1997-2001, J. Geophys. Res., 107(D17),   4337.

Long-term measurements of CO, NO, NO2, organic compounds and PM10 at a motorway location in an Austrian valley

R. Schnitzhofer[1], J. Dunkl[1], J. Beauchamp[1], A. Weber[2], A. Wisthaler[1] and
A. Hansel[1]

[1]*Institut für Ionenphysik, Leopold Franzens Universität Innsbruck, 6020 Innsbruck,
Austria (ralf.schnitzhofer@uibk.ac.at)*

[2]*Landesforstdirektion Waldschutz/Luftgüte, Amt der Tiroler Landesregierung, 6020
Innsbruck, Austria*

ABSTRACT

In situ measurements of CO, NO_x, PM_{10} and certain organic compounds within the River Inn valley (Tirol, Austria) have been taking place since February 2004. The monitoring site experiences varying meteorological conditions and traffic frequency throughout the day, both of which strongly influence local air pollutant levels. Large increases of NO_x in the early morning are clearly correlated with increasing heavy duty vehicle (HDV) abundance, whereas concentrations of certain volatile organic compounds (VOCs), such as benzene and toluene, correspond to an increase in light duty vehicle (LDV) numbers. After a morning peak, concentrations of air pollutants decrease due to vertical dilution of the valley air, reaching a minimum at about midday when the planetary boundary layer is well mixed. Afterwards there is a slight increase in concentrations when the boundary layer slowly returns to stable conditions. This daily undulation is seasonally dependent, varying slightly throughout the year due to the influence of sunrise on the stability of the valley atmosphere. The seasonal variation of morning peak concentrations indicate smallest maxima occurring in June (NO concentration of about 350 µg m^{-3}), correlating with an early sunrise. Conversely, highest morning concentrations (NO reaching 550 µg m^{-3}) arise in early spring and autumn due to a late, reduced boundary layer dilution.

1 Introduction

There is currently much debate about the effects of vehicle exhaust on local air quality. Many volatile organic compounds (VOCs) can have serious repercussions on human health, both directly and indirectly. Benzene for example is known to be a carcinogenic agent (WHO, 2000), whereas other VOCs lead to O_3 production in the presence of NO_x (Staffelbach & Neftel, 1997), with O_3 known to be a skin irritant and damaging to vegetation. In the past 100 years the oxidation of VOCs under these conditions has led to a significant increase of O_3 in the planetary boundary layer (Volz & Kley, 1998).

Air quality problems are particularly important in the Alps, where the mountainous landscape prevents horizontal dilution on the one hand, and on the other hand generates special meteorological conditions that restrict vertical dilution. Such conditions include calmer wind situations and longer-lasting inversion layers than in flat landscapes, both of which limit dilution. Furthermore, recirculation of pollutants within a valley wind system contributes to higher concentrations, as well as limiting dilution-volume in a valley. Wotawa et al. (2002), for instance, found that morning NO_x concentrations in winter are up to a factor 9 higher in the River Inn valley compared to flat terrain with the same amount of traffic.

In order to monitor air quality in Tirol, a number of measurement sites routinely monitoring certain parameters are in operation. One of these sites is located alongside the A-12 motorway in Vomp within the River Inn valley (Tirol, Austria), where the present investigation takes place. In a previous short-term study at this location in autumn 2002, Beauchamp et al. (2004) monitored VOCs to complement routine data acquisition of NO, NO_2, CO and PM_{10}. Increasing NO_x levels were found to be predominantly attributable to heavy duty vehicle (HDV) traffic.

The current investigation began in February 2004 and is still running. The aim is to study the long-term situation at this location.

2. Experimental

The monitoring site in Vomp is situated no more than two meters from the offside of the A-12 motorway hard shoulder, making it ideal for immediate vehicle emission detection. A proton-transfer-reaction mass spectrometer (PTR-MS; described in detail by Hansel et al., 1995) was set up at the monitoring location in February 2004 (measurements still ongoing at present) to detect key organic compounds (e.g. methanol, acetone, benzene, toluene, etc.) in the air in the vicinity of the motorway. The heated ca. 4 m inlet line, elevated approximately 4 m above ground level and capped with a particulate filter, leads ambient air at a constant flow of 2 l/min to the PTR-MS. Each compound is measured on a continuous mass scan cycle from atomic mass 20 to 200 amu, with each cycle lasting approximately 6 min. Continuous measurements of NO, NO_2, CO and PM_{10} are additionally routinely made at this site using standard commercial instrumentation. Traffic numbers in the vicinity of the monitoring site are consistently measured by means of induction bands across the carriageway and are separable into HDV and light duty vehicles (LDV).

To gain understanding of the meteorological conditions in the valley, wind direction and speed are monitored at the measurement site at an elevation of approximately 6 m. Additionally, temperature is monitored at 8 different heights on the Kellerjoch Mountain, which is situated several kilometres southeast of the motorway site on the opposite side of the valley. Temperature profile data are used to approximate the vertical dilution conditions in the valley.

3. Results and Discussion

Motorway driving restrictions on HDV in Tirol allowed data to be separated into HDV-present and HDV-absent periods. Figure 1 shows monthly average diurnal variations of NO on days when HDV were allowed to use the motorway (from 05:00 to 22:00, local time). Data from March through October 2004 are plotted, clearly displaying seasonal differences.

In the morning a sudden increase of NO strongly correlates to increasing HDV numbers (not displayed; see fig. 2 for HDV abundance in April 2004). NO concentrations peak between 06:30 and 08:00 local time. The time and size of these maxima correlates with sunrise, so that the latest and largest maxima occur in March and October, with NO concentrations of between 500 and 550 $\mu g\ m^{-3}$. The earliest and smallest maximum, when early sunrise leads to early vertical dilution, occurs in June, where NO concentrations reach only 350 $\mu g\ m^{-3}$. At about midday, concentrations reach a minimum when the planetary boundary layer is well mixed. The deepest midday minima occurring in March and April may be due to a large temperature difference between the snow-free valley floor and the snow-covered slopes, which results in well developed convection. Secondary maxima occur in the evening, when the planetary boundary layer slowly returns to stable conditions. NO levels decrease again at 22:00 local time when HDV are once again restricted from using the motorway. The above described diurnal variation of NO does not occur on Sundays when HDV are banned for the whole day. Instead, NO concentrations remain low (data not shown).

Figure 2 shows average volume mixing ratios (VMRs) of benzene and toluene on working days in April, as well as average HDV and LDV numbers. Benzene and toluene show a morning increase correlating with a rise in LDV numbers. A strong decrease begins to occur at about 08:00 local time, despite vehicle numbers remaining high. This is attributable to the gradual dispersion of the lowest inversion layer, leading to the development of a well mixed planetary boundary layer over the next few hours.

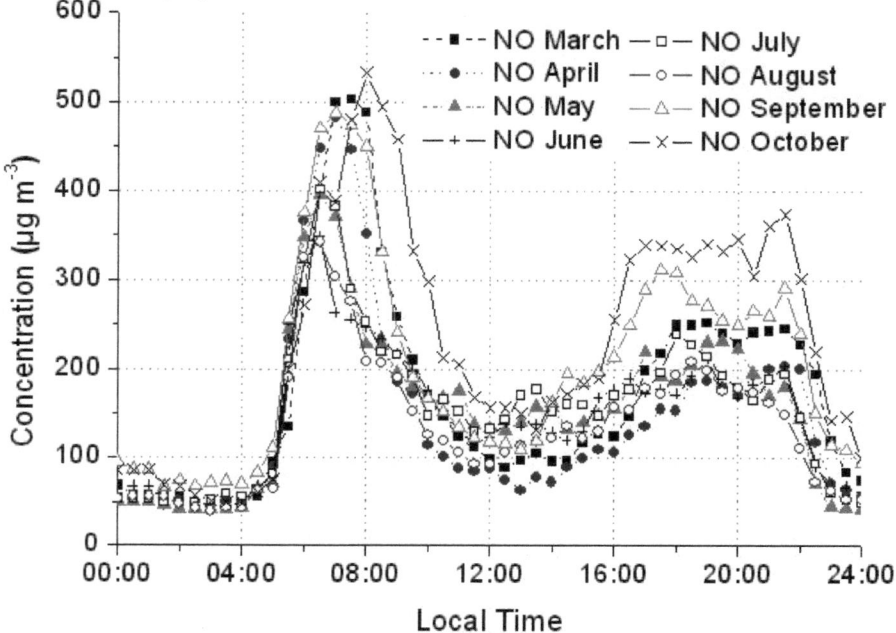

Figure 1: Concentration of NO (30 min. means, in μg m⁻³) for March through October 2004, averaged over work days when HDV are allowed to use the motorway from 05:00 to 22:00 local time.

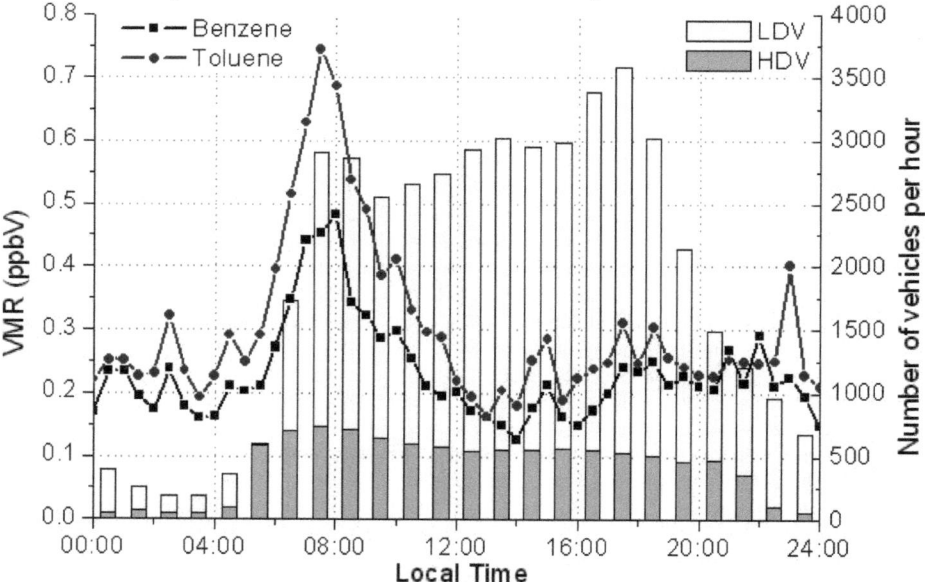

Figure 2: VMRs of benzene and toluene (30 min. means, in ppbv), and numbers of HDV and LDV (in vehicles per hour) averaged over working days in April 2004.

4. Conclusions

Preliminary results from this ongoing investigation indicate a strong dependence of meteorology and traffic abundance on the levels of the different compounds being monitored. NO concentrations clearly correlate with HDV abundance, whereas VMRs of the aromatic compounds show variations with LDV numbers. Additionally, both classes of compounds are dependent on local meteorological conditions: During early spring and autumn maxima are higher and occur later due to a late sunrise and the large temperature gradients present from the snow-covered mountain slopes to the snow-less valley floor. A seasonally varying restriction on HDV using the motorway would likely help reduce these higher maxima. CO and PM$_{10}$ data are currently in the evaluation phase. Preliminary results will be presented.

Acknowledgements

The authors would like to thank the Amt der Tiroler Landesregierung, Landesbaudirektion, Abteilung Gesamtverkehrsplanung, in particular E. Allinger-Csollich for the vehicle number data collection. Additional thanks go to the Brenner Eisenbahngesellschaft and the NÖ Umweltschutzanstalt, Abteilung Luftreinhaltung, Maria Enzersdorf, in particular A. Amann, for temperature data from the Kellerjoch Mountain in Schwaz.

References

Beauchamp, J., A. Wisthaler, W. Grabmer, C. Neuner, A. Weber, A. Hansel **(2004)**; *Short-term measurements of CO, NO, NO2, organic compounds and PM10 at a motorway location in an Austrian valley*; Atmospheric Environment 38, 2511-2522.

Hansel, A., A. Jordan, R. Holzinger, P. Prazeller, W. Vogel, W. Lindinger **(1995)**; *Proton transfer reaction mass spectrometry: on-line trace gas analysis at ppb level*; International Journal of Mass Spectrometry and Ion Processes 149/150, 609-619.

Staffelbach, T. and A. Neftel **(1997)**; *Relevance of biologically emitted trace gases for the ozone production in the planetary boundary layer in Central Europe*; Schriftenreihe der FAL 25.

Volz, T. and D. Kley **(1988)**; *Evaluation of the Montsouris series of ozone measurements made in the nineteenth century*; Nature 332, 240–242.

WHO (2000); *Air Quality Guidelines for Europe, 2nd Edition*; World Health Organisation, Regional Office for Europe, Copenhagen. WHO Regional Publications, 91.

Wotawa, G., P. Seibert, H. Kromp-Kolb, M. Hirschberg **(2002)**; *Verkehrsbedingte Stickoxid-Belastung im Inntal: Einfluss meteorologischer und topographischer Faktoren, Endbericht zum Projekt Nr. 6983*; Inst. für Met. und Physik, Universität für Bodenkultur Wien; Oktober 2002.

PTR-MS response-time improvements

G. Hanel[1], W. Sailer[2], and A. Jordan[1]

[1] Ionicon Analytik GesmbH. Technikerstrasse 21a, 6020 Innsbruck, AUSTRIA

[2] Ionimed Analytik GesmbH. Technikerstrasse 21a, 6020 Innsbruck, AUSTRIA ...

One big advantage of PTR-MS is the very fast response time. Especially in fast dynamic systems like flavour analysis or atmospheric flux measurements each small time improvement is directly connected to the accuracy of the measurements. Although it is necessary to have higher concentrations to decrease the integration time for every mass it is now possible to collect the measured data approximately one order of magnitude faster than before. Using a new designed inlet system and a completely new developed software package we are now able to achieve a response time of down to 85ms in the increasing signal part and a response time in range of 85 to 120ms in the decreasing signal (see Fig. 1).

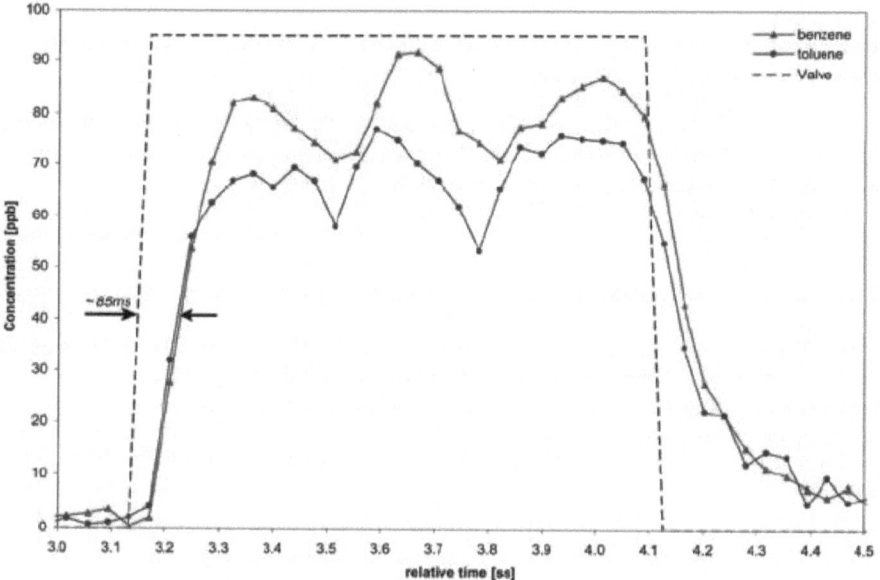

Fig. 1: Illustration of response time when opening an inlet valve for one second

The main improvement besides the new inlet design was the new software design. This speed improvement is only practicable when using the newly developed PTR-MS software package. This package is a replacement or parallel usable software for the up to now used workplace software. Data recorded using the PTR-MS control 2005 software is stored directly in ascii formatted spreadsheet file format. The newly developed communication engine communicating between the computer and the quadrupole mass spectrometer is optimized for the maximum throughput of the RS232 communication line. So the maximum achievable data rate directly transferred into the PC and not using the quadrupoles build in memory is 250Hz. In the moment this new software package is in beta state. It can be obtained from Ionicon Analytik on request. Due to the beta state service and support of this version is very limited.

Index of Authors

Talbot R.W. 28
Tani A. 23
Telser S. 98
Tilg H. 97
Tillmann R. 41
Tonini A. 73, 134
Tschiersch J. 121

V

van Ruth S. 81, 112, 130
van Swam K. 96
Versini G. 148
Verucchi R. 73, 134
Vescovi M. 73, 134
Viidanoja J. 153

W

Wagner A. 151
Wargocki P. 60
Warneke C. 52
Weber A. 73, 134, 166
Weller U. 158
Weschler C.J. 60
Wildt J. 158
Wilkinson M. 47
Wisthaler A. 19, 60, 150, 151, 152, 153, 154, 158, 166
Wyon D.P. 60

Z

Zahn A. 162
Zini E. 138